Tribological Analysis and Design of a Modern Automobile Cam and Follower

ENGINEERING RESEARCH SERIES

Tribological Analysis and Design of a Modern Automobile Cam and Follower

G Zhu and C M Taylor

Series Editor
Duncan Dowson

Professional Engineering Publishing

Professional Engineering Publishing Limited,
London and Bury St Edmunds, UK

First published 2001

ISBN 978-1-86058-203-5

ISSN 1468-3938
ERS 7

A CIP catalogue record for this book is available from the British Library.

Related Titles

IMechE Engineers' Data Book – Second Edition	C Matthews	ISBN 1 86058 248 6
Design Techniques for Engine Manifolds – Wave Action Methods for IC Engines	D E Winterbone and R Pearson	ISBN 1 86058 179 X
Theory of Engine Manifold Design – Wave Action Methods for IC Engines	D E Winterbone and R Pearson	ISBN 1 86058 209 5
Handbook of Automotive Powertrain and Chassis Design	J Fenton	ISBN 1 86058 075 0
Statistics for Engine Optimization	Eds S P Edwards, D M Grove, and H P Wynn	ISBN 1 86058 201 X
Integrated Powertrains and their Control	Ed. N Vaughan	ISBN 1 86058 334 2
Journal of Automobile Engineering	Part D of the Proceedings of the IMechE	ISSN 0954–4070

Other titles in the Engineering Research Series

Industrial Application of Environmentally Conscious Design (ERS 1)	T C McAloone	ISBN 1 86058 239 7 ISSN 1468–3938
Surface Inspection Techniques – Using the Integration of Innovative Machine Vision and Graphical Modelling Techniques (ERS 2)	M L Smith	ISBN 1 86058 292 3 ISSN 1468–3938
Laser Modification of the Wettability Characteristics of Engineering Materials (ERS 3)	J Lawrence and L Li	ISBN 1 86058 293 1 ISSN 1468–3938
Fatigue and Fracture Mechanics of Offshore Structures (ERS 4)	L S Etube	ISBN 1 86058 312 1 ISSN 1468–3938
Adaptive Neural Control of Walking Robots (ERS 5)	M J Randall	ISBN 1 86058 294 X ISSN 1468–3938
Strategies for Collective Minimalist Mobile Robots (ERS 6)	C Melhuish	ISBN 1 86058 318 0 ISSN 1468–3938

For the full range of titles published by Professional Engineering Publishing contact:

Sales Department
Professional Engineering Publishing Limited
Northgate Avenue
Bury St Edmunds
Suffolk
IP32 6BW
UK

Tel: +44 (0)1284 724384
Fax: +44 (0)1284 718692
www.pepublishing.com

Contents

Series Editor's Foreword

The nature of engineering research is such that many readers of papers in learned society journals wish to know more about the full story and background to the work reported. In some disciplines this is accommodated when the thesis or engineering report is published in book or monograph form – describing the research in much more complete form than is possible in journal papers. The Engineering Research Series offers this opportunity to engineers in universities and industry, and thus disseminates wider accounts of engineering research progress than are currently available. The volumes supplement and do not compete with the publication of peer-reviewed papers in journals.

The volumes selected for publication usually arise from one of the following sources; single higher degree theses, a series of theses on a particular engineering topic, submissions for higher doctorates, reports to sponsors of research, or comprehensive industrial research reports. Furthermore, since it is quite usual for university engineering research groups to undertake research on problems reflecting their expertise over several years, it may be appropriate to produce a comprehensive account of the development of understanding and knowledge on a selected topic in a specially prepared single volume. The Series Editor would be pleased to discuss such proposals.

Factors to be considered in the selection of items for the Series include the intrinsic quality of the volume, its comprehensive nature, the novelty of the subject, potential applications, and the relevance to the wider engineering community.

Volumes have already been accepted in the following areas:

Environmentally conscious design integration
Surface inspection techniques
Laser modification of the wettability characteristics of engineering materials
Fatigue and fracture mechanics of offshore structures
Adaptive neural control of walking robots
Strategies for collective minimalist mobile robots

Authors are invited to discuss ideas for new volumes with Sheril Leich, Commissioning Editor in the Books Department, Professional Engineering Publishing Limited, or with the Series Editor directly.

The present volume, which is the seventh to be published in the Series, comes from the University of Leeds and is entitled

Tribological Analysis and Design of a Modern Automobile Cam and Follower
by
Dr G. Zhu and Professor C. M. Taylor

The work on which this volume is based was carried out in the School of Mechanical Engineering in the University of Leeds, under sponsorship from the Ford Motor Company. It represents a fine example of collaboration between the University and industry, in which tribological expertise in the University was utilized in an approach to a problem of considerable significance to a major design and manufacturing company.

The automobile engine valve train presents a major challenge to the tribologist. The severe dynamic loading on the cam–follower mechanism is often transmitted between components of uncompromising form. The kinematics of the functional cycle fails to support consistent lubricant entraining motion and squeeze film action merely ameliorates the severity of mixed-film lubrication during part of the cycle. Few highly taxed, lubricated machine components present a greater challenge to the tribologist.

The primary objective of the analysis reported is to apply state-of-the-art tribological knowledge to the overall prediction of friction and power loss in complete valve trains. There are three main sections: theoretical formulation; an experimental study; and a correlation of theory and experimental results.

The work will be of particular interest to designers and manufacturers of reciprocating engines, valve train specialists, and tribologists. It represents a valuable addition to the Engineering Research Series of books.

Professor Duncan Dowson
Series Editor
Engineering Research Series

Authors' Preface

Research studies in the area of lubrication engineering, now known as tribology, have flourished in the School of Mechanical Engineering at the University of Leeds for over a century. Such studies have embraced fundamental considerations, strategic developments, and practical applications to machine elements such as belts, cams and followers, dry bearings, gas bearings, gears, hydrostatic bearings, plain bearings, piston rings, rolling element bearings, and seals.

Research at Leeds into the major frictional components associated with the internal combustion engine – the piston assembly, valve train, and bearings – began some 30 years ago. In the early 1970s Professor Duncan Dowson began what has been a 30-year period researching into the predicted and measured performance of piston assemblies, and it was also in the 1970s that studies of engine bearing technology first commenced. In the early 1980s, Duncan Dowson and Chris Taylor approached the Ford Motor Company with the suggestion that it was timely to begin analysis of the predictions of engine friction as a whole. This was well received by Ford, who had supported research work at Leeds over a number of years. It was recognized that it would be appropriate in the first instance to address the main frictional components of the internal combustion engine and leave studies of friction associated with accessories to a later time. It was, however, soon appreciated that while there was substantive understanding of the performance of various valve train mechanisms, the ability to undertake detailed design analysis and predictions from a tribological perspective was not strong. This initiated a programme of research extending over some 10 years at the University of Leeds, supported by the Ford Motor Company.

The first research programme at Leeds was undertaken by Paul Harrison who received his doctorate in 1985 for a thesis 'A study of the lubrication of automotive cams'. This research addressed a direct acting cam and follower (the cam and bucket follower), which at that time was not a common form of valve train mechanism in the United Kingdom. It has since moved to a position of prominence in terms of application. The work was continued by Guangrui Zhu in his thesis 'A theoretical and experimental study of the tribology of a cam and follower', extending the initial analysis undertaken by Dr Harrison, and by Andrew Ball who studied a range of valve train mechanisms in his work 'The tribological study of the design and performance of automotive cams'. Both these theses were published in 1988. Interestingly enough, the work on developing an engine friction model with Ford was taken forward subsequent to these research programmes and Li Sheng Yang presented his thesis 'Friction modelling for internal combustion engines' in 1992.

All the above research programmes were generously supported by the Ford Motor Company, reflecting their continuing commitment to addressing leading technological developments in support of their products. Not only was the detailed understanding of the tribological performance of a variety of cam and follower mechanisms developed (including work by the Industrial Unit of Tribology at Leeds on more practical aspects) but a world-leading friction model for automobile engines was delivered to Ford in the early 1990s.

Subsequent to completion of his PhD thesis, Guangrui Zhu was supported as a postdoctoral Research Fellow by the Ford Motor Company, with the support of Chris Taylor, to develop a theoretical and experimental study of the direct acting cam and follower (with an hydraulic adjuster) in the Ford ZETEC engine. It is the report on that work which forms the basis of this book. Throughout the partnership between the Institute of Tribology and the Ford Motor Company over many years there has been tremendous support from a variety of colleagues at Ford in relation to managing the projects, procuring engine parts, providing technical data and, of course, generous financial support. Particular thanks are, however, due to Mr Sasan Dadgostar and Mr Nigel Weaver for their substantive contributions.

G Zhu
Cummins Engine Company Limited
Daventry
United Kingdom

C M Taylor
Institute of Tribology
School of Mechanical Engineering
The University of Leeds
United Kingdom

Notation

a	semi-major axis of ellipse of the contact
A	apparent area of contact
A_o	area of the orifice of the check valve
b	semi-minor axis of ellipse of contact
B_1	length of lower section of follower
B_2	length of upper section of follower
C_{b1}	minimum separation between follower and bush (bottom)
C_{b2}	minimum separation between follower and bush (top)
C_f	radial clearance between follower and bore
C_v	radial clearance between valve stem and guide
C_w	cam width
D_f	diameter of follower
D_v	diameter of valve stem
e	eccentricity of point of contact from camshaft centre
e_f	offset of cam to centre plane of follower
e_1	eccentricity of bottom half of follower
e_2	eccentricity of top half of follower
E_c	Young's modulus of cam material
E_f	Young's modulus of follower material
E'	equivalent elastic constant
F_c	friction of cam/follower interface
F_f	friction of follower/bore interface
F_v	friction of valve/guide interface
G	dimensionless materials parameter, $\alpha E'$
h	nominal separation between two reference surfaces
h_{cen}	central lubricant film thickness
$h_{\min}$	minimum lubricant film thickness
h_1	radial clearance between outer surface of plunger and bush
h_2	radial clearance between inner surface of plunger and cylinder
H	power loss
H_1	overlap between outer surface of plunger and bush
H_2	overlap between inner surface of plunger and cylinder
J	moment of inertia of follower
k	ellipticity ratio of semi-major axis to semi-minor axis, *a/b*
k_{ps}	stiffness of plunger spring
k_{vs}	stiffness of valve spring
l_c	cam lift
l_f	follower lift
$\dot{l}_f$	velocity of follower
$\ddot{l}_f$	acceleration of follower
L_f	length of follower
L_v	length of valve guide

m_f	mass of follower
m_p	mass of plunger
m_s	mass of valve spring
p	local pressure
p_o	oil pressure in follower reservoir
p_1	oil pressure in plunger chamber
$p_{\max}$	maximum Hertzian pressure
r_o	radius of orifice of check valve
r_1	radius of plunger
r_2	radius of plunger cylinder
R_b	radius of cam base circle
R_e	effective radius of curvature in direction of lubricant entrainment
R_f	radius of curvature of domed follower surface
R_p	radius of curvature of plunger end
R_s	effective radius of curvature in direction of side-leakage
R_x	equivalent radius of curvature in x direction
R_y	equivalent radius of curvature in y direction
U_e	dimensionless speed parameter, $\eta_o u_e/(E'R_e)$
W	load on cam
W_a	asperity load
W_e	dimensionless load parameter, $W_h/(E'R_e^2)$
W_h	hydrodynamic load, $W - W_a$
x, y	co-ordinate system
z	plunger displacement relative to follower
α	viscosity–pressure coefficient of lubricant
α'	cam taper
β	asperity radius of curvature or cam back taper
γ	rate of change of shear stress with pressure
δ	precompression of valve spring
δ_p	precompression of plunger spring
δ_t	shift of first contact point from central plane of follower
ε_1	dimensionless eccentricity of bottom half follower
ε_2	dimensionless eccentricity of top half follower
ζ	follower tilting angle
η	dynamic viscosity of lubricant or asperity density
η_b	dynamic viscosity of lubricant within tappet/bore interface
η_o	dynamic viscosity of lubricant at atmospheric pressure
θ	follower angular displacement
λ_{b1}	ratio of separation between tappet and bore (bottom)
λ_{b2}	ratio of separation between tappet and bore (top)
μ_a	asperity sliding coefficient of friction
μ_b	boundary lubrication coefficient of friction
μ_c	friction coefficient of cam/follower interface
υ_c	Poisson's ratio of cam material
υ_f	Poisson's ratio of follower material
ξ	coefficient of flow loss of orifice of check valve

ρ	density of lubricant
σ	composite roughness height of cam/follower interface, $\sqrt{\sigma_c^2+\sigma_f^2}$
σ_{fb}	composite roughness height of follower/bore interface
τ_o	Eyring stress of lubricant
τ_L	limiting shear stress of lubricant
ϕ	cam angle
ω	rotational frequency of camshaft
Ω	rotational frequency of follower

Subscripts

a	asperity
b	base circle of the cam or bore of follower
c	cam
f	follower
p	plunger
s	seal
v	valve

Other variables may be defined in the text where necessary.

Chapter 1

Introduction

Since the 1973 Arab oil embargo, reduction in fuel consumption has become an important issue. Other reasons for improving fuel economy are: pollution, running cost (fuel is now more than 80 pence/litre), and conservation. Emission control and fuel economy have had a great influence on recent engine designs. The general trend in engine design has been towards the improvement of combustion. The adoption of double intake and exhaust valves (16 valves for a four-cylinder engine) is one of the alternatives. With a carefully designed intake manifold, the two intake valves create two separate streams of inlet air swirling in opposite directions to improve the breathing and the combustion of the engine.

On the other hand, in order to improve the breathing of the engine, cam opening rates can be increased. This requires valve train stiffness of a higher order than can be achieved by a pushrod valve train. Thus the overhead cam valve train came into favour. This comes in many forms; 'bucket follower', 'rocker follower', and 'finger follower' being the main categories. The bucket follower is a relatively low-friction system, which also has low rates of wear. For a four-cylinder engine with 16 valves, two camshafts are needed if bucket followers are to be used. This means the power loss arising from the valve train will be doubled.

The Institute of Tribology at the University of Leeds has for some time collaborated with the Ford Motor Company Limited on studies of automobile valve trains and wider aspects of engine friction. The valve train studies have covered all forms of automotive cam and follower mechanisms and entailed experimental and analytical research. A wide-ranging development of design software for the prediction of the tribological performance of all valve train forms has been undertaken by the Institute and has been installed on Ford computers. In addition to the analytical studies comprehensive experimental programmes have been conducted to investigate the tribological performance of a cam and flat-faced (bucket) follower. Several important factors including the friction torque, power loss, electrical resistivity across the contact of the cam and follower, and temperatures at the cam nose and in the valve guide were monitored in each test.

At the present time there is a strong trend towards the use of roller followers and bucket followers. With the drive for quieter engines and longer service intervals the hydraulic lash adjuster is also becoming more and more common. While much of the information presented by the Institute has related to cam and flat-faced follower with nominal line contact, there remain many challenging problems in more practical arrangements. An analysis to understand the tribological implications of the arrangement having an offset taper cam and a domed rotating follower incorporated with an hydraulic lash adjuster is complex and challenging.

As a research project supported by the Ford Motor Company, the work presented in this book has been concerned with both theoretical and experimental studies of the tribological performances of an automobile valve train having an offset taper cam and a domed rotating follower incorporated with an hydraulic lash adjuster with particular reference to the ZETEC engine valve train. The arrangement of the valve train is shown in Fig. 1.1. The work has included three main parts: (a) the theoretical formulation; (b) the experimental investigation and validation; and (c) the correlation of theory and experiments.

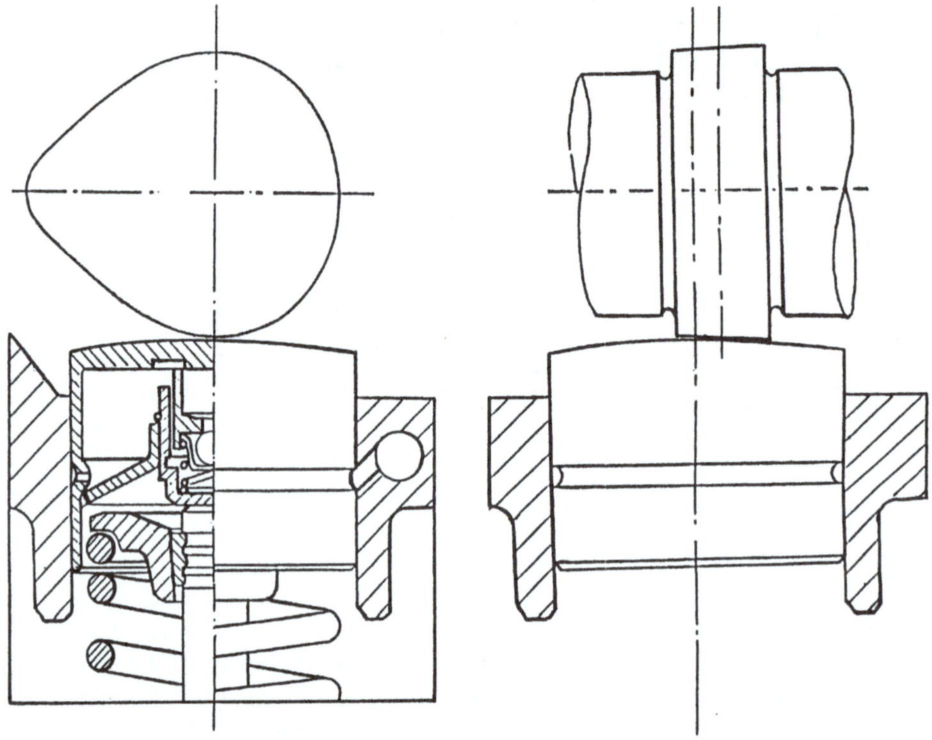

Fig 1.1 Domed tappet incorporating hydraulic lash adjuster

First, a sophisticated theoretical model has been developed to predict the tribological performance of the valve train. The development of the model is to provide a useful tool for the tribological design of the engine valve train. Several important tribological characteristics of the valve train, such as the maximum Hertzian stress, the central and minimum lubricant thickness, and the friction and power loss arising from the interfaces of the valve train including cam/follower, tappet/bore, and valve stem/guide interfaces, can be predicted by this model. In addition to this, the model can estimate the instantaneous and the average rotational frequency of the follower. Furthermore, the performance of the hydraulic lash adjuster can be predicted by the current theoretical model.

Second, a careful and comprehensive experimental programme consisting of a series of tests at different camshaft rotational frequencies and different bulk temperatures has been undertaken. This is to investigate the tribological performance of the valve train and also to validate the theoretical model. Several important variables including the friction torque, the power loss, the follower rotational frequency, and the temperatures of the lubricant supply, inside the test cell and in the valve guide, have been monitored in each test. Some sophisticated techniques for data acquisition and data processing, including a PC-based data acquisition system and a follower rotational frequency monitoring system have been developed. The accomplishment of the comprehensive experimental programme has provided a vast amount of reliable data. This enables the validity of the theoretical model to be examined.

Finally, in order to validate the theoretical model, the experimental measurements have been correlated with the theoretical predictions that simulate the test conditions of the valve train. The agreement between the experimental measurements and the theoretical predictions is excellent.

Part I – Theoretical Formulation

Chapter 2

Kinematics and Dynamics of the Cam and Follower

2.1 Introduction

In the first part of the book detailed introductions have been given to the theoretical formulation of the model. The theoretical basis of the model and the solution procedures of the computer program coded in FORTRAN 77 have been illustrated in the following seven chapters. At the end of each chapter a list of references relating to the development of the current theoretical model is provided. Then an introduction to the overall construction of the program and arrangements of the input and output data of the program have been listed in Chapter 9. Finally an example of the tribological analysis of the valve train with particular reference to the design case of the ZETEC engine valve is demonstrated with a set of graphics. A brief discussion of the results is also undertaken.

In this chapter the kinematic and dynamic analyses of the valve train having an offset taper cam and a domed follower are outlined. In this book equations and figures are numbered with two digits separated by a dot. The first number indicates the chapter in which the equation or the figure first appears and the second one represents the sequence. For example, equation (8.5) represents an equation outlined in Chapter 8 with a sequence number (5).

2.2 Kinematic analysis of a cam and domed follower pair

The kinematic analysis provides information about the velocity of the contact point relative to the cam and follower [1]. The velocity of the point of contact relative to the cam, V_c, is given by

$$V_c = \omega \left[L^2 + \left(\frac{dL}{d\phi}\right)^2 - R_f \frac{L^2 + \left(\frac{dL}{d\phi}\right)^2 - L\frac{d^2L}{d\phi^2}}{L^2 + \left(\frac{dL}{d\phi}\right)^2} \right] \tag{2.1}$$

where
R_f is the radius of curvature of domed follower surface;
ω is the camshaft rotational frequency;
ϕ is the cam angle.

The velocity of the point of contact relative to the follower V_{fm} by assuming the follower is stationary is given by

$$V_{fm} = \omega R_f \frac{L\dfrac{d^2L}{d\phi^2} - \left(\dfrac{dL}{d\phi}\right)^2}{L^2 + \left(\dfrac{dL}{d\phi}\right)^2} \tag{2.2}$$

The cam radius of curvature at the contact point R_c is calculated from

$$R_c = \frac{\left[L^2 + \left(\dfrac{dL}{d\phi}\right)^2\right]^{\frac{3}{2}}}{L^2 + 2\left(\dfrac{dL}{d\phi}\right)^2 - L\dfrac{d^2L}{d\phi^2}} - R_f \tag{2.3}$$

where

$$L = l_c + R_b + R_f \tag{2.4}$$

l_c is the cam lift;
R_b is the radius of the cam base circle.

If the follower is rotating, the velocity V_f of the contact point relative to the follower will be modified. Figure 2.1 shows a plan view of the follower surface. The axis of rotation of the follower is at O'. An xoy co-ordinate system is adopted with the origin of the co-ordinate fixed at O, the centre of the contact area as shown in the figure. The rotation of the follower introduces an additional velocity component V_{fr} of the contact point relative to the follower. At a given point, A, in the contact area, V_{fr} is equal to the product of the follower rotational frequency Ω and the distance d between A and O'. The direction of the velocity is perpendicular to the line AO'. Thus the resultant velocity of the contact point relative to the follower should be written as

$$\begin{aligned} V_f &= \left|\vec{V}_{fm} + \vec{V}_{fr}\right| = \sqrt{U_{fx}^2 + U_{fy}^2} \\ U_{fx} &= V_{fm} + \Omega(\delta_t + y) \\ U_{fy} &= \Omega(e - x) \\ \delta_t &= R_f \sin\alpha' \end{aligned} \tag{2.5}$$

where
U_{fx} is the velocity component in the x direction;
U_{fy} is the velocity component in the y direction;
Ω is the rotational frequency of the follower;
α' is the cam taper;
e_f is the offset of the cam to the centre of the follower;
e is the eccentricity of the point of the contact from the centre of cam rotation.

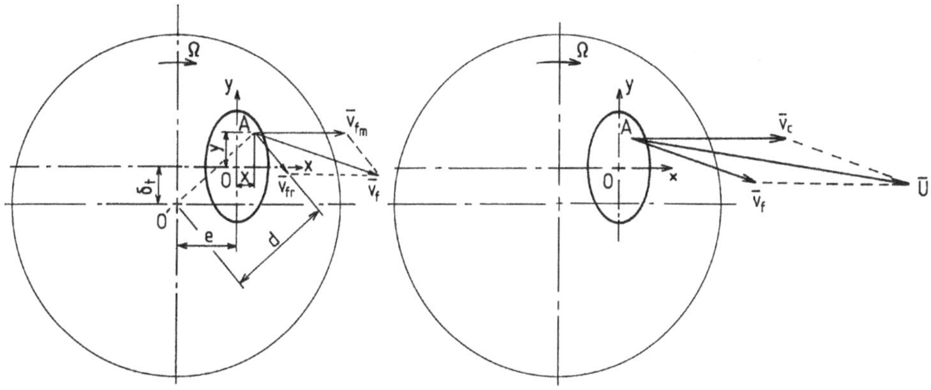

Fig. 2.1 Plan view showing contact area projected on to the follower

2.3 The load at the cam/follower interface

The force associated with a cam and follower pair needs to be known from both the lubrication and stress points of view. In general, the forces associated with the operation of a cam mechanism are: the inertia force, the spring force, the dynamic deflection and damping effect associated with the parts, and the friction between the moving parts. In the present analysis the assumption of a rigid mechanism was made, the load between the cam and follower being concentrated in the centre of the contact O. Figure 2.2 shows the balance of the forces acting on the domed follower. The friction F_c between the cam and follower is taken to be equal to μW, where μ is the friction coefficient between the cam and follower. Projecting the forces in the z direction leads to the load on cam of

$$W = \frac{P + F_f + m_f\left(\ddot{l}_f + \ddot{z}\right)}{\cos\beta + \mu\sin\alpha}$$

$$\alpha = \sin^{-1}\left(\frac{e}{\sqrt{R_f^2 - \delta_t^2}}\right) \qquad (2.6)$$

$$\beta = \sin^{-1}\left(\frac{\sqrt{\delta_t^2 + e^2}}{R_f}\right)$$

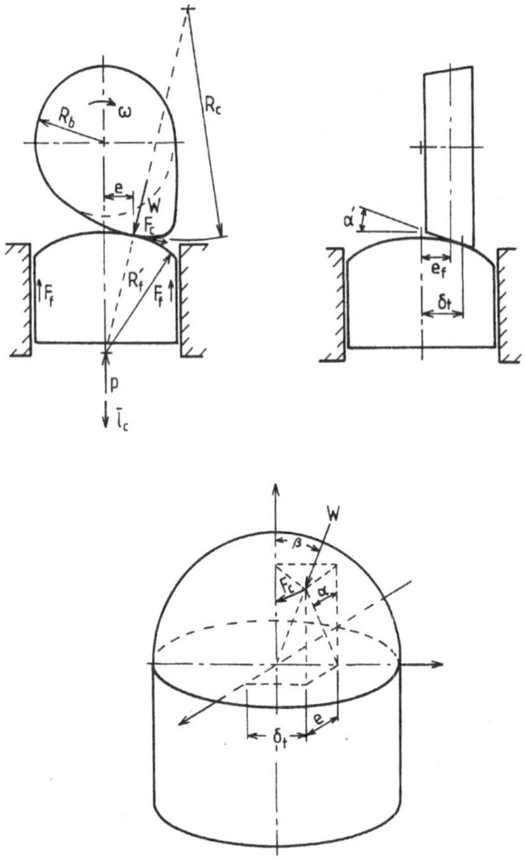

Fig. 2.2 Loading on the cam and follower

where

m_f is the mass of the follower.

$\ddot{l}_f$ is the acceleration of the follower. If the radius of the inspection cylinder for the cam lift is equal to the radius of the follower surface, the follower lift l_f is equal to the cam lift l_c. Thus, the follower velocity $\dot{l}_f$, and acceleration $\ddot{l}_f$, are equal to $\dot{l}_c$ and $\ddot{l}_c$ respectively, and can be calculated with numerical differentiation of l_c with respect to time. In the same way, $\ddot{z}$ can be obtained from z.

P is the force applied by the valve stem and is given by equation (3.5).

F_f is the friction arising from the interface of the follower and its bush and is given in Section 7.3.

On the cam base circle the valve stops moving. Thus the load remaining on the cam is only due to the plunger spring and the oil pressure applied on the plunger. The oil pressure in the plunger chamber is assumed to be equal to the pressure in the follower reservoir p_o. So the load on the cam base circle is given by

$$W = k_{ps}(\delta_p - z) + Ap_o \tag{2.7}$$

where
k_{ps} is the stiffness of the plunger spring;
δ_p is the precompression of the plunger spring;
p_o is the oil pressure in the follower reservoir;
A is the inner bottom area of the plunger (see Section 3.2.1, list point 3 and Fig. 3.2).

2.4 The solution procedure

The solution procedure of the kinematic and dynamic analysis is in the sequence introduced in this chapter. There are three unknowns in the above analysis. They are the friction coefficient between the cam and follower μ, the follower rotational frequency Ω and the plunger displacement z. These unknowns are only available from the analyses on the friction (Chapter 7), follower rotation (Chapter 8) and the hydraulic lash adjuster (Chapter 3) later. So, starting with three initial values, the analyses of the kinematics and dynamics, the stresses, the hydraulic lash adjuster, the rotation of the follower, the oil film thickness, and then the friction are performed. The real values of these three unknowns are determined from three iterative loops which are shown in the flow chart of Fig. 9.1.

2.5 References

1. **Ball, A.** (1988) A tribological study of the design and performance of automotive cams. PhD thesis, Mechanical Engineering Department, Leeds University, UK, November.

Chapter 3

Hydraulic Lash Adjuster

3.1 Introduction

In this chapter a theoretical model is developed to predict the plunger movement relative to the follower. This is essential in considerations of the dynamics of the valve train of the ZETEC engine and is also important to the tribological analysis of the valve train. In the present analysis it was assumed that the hydraulic lash adjuster worked with either the check valve open or the check valve closed. The theoretical model was developed separately in respect of each condition.

3.2 Check valve closed

During most of the cam event, the hydraulic lash operates under a loaded condition with the check valve closed. The movement of the plunger is caused by its external loading which squeezes out the oil in the plunger chamber through the clearance between the plunger and inner cylinder. As the amount of escaping oil is very small, the plunger displacement relative to the follower is expected to be far less than the cam lift. This means that the movement of the follower, and consequently the valve, follows the cam lift (or the follower is pressured on to the cam by the valve spring and the inertia of the valve train).

3.2.1 The dynamics of the plunger

The movement of the plunger relative to the follower depends upon the forces associated with the plunger. Figure 3.1 shows a layout of the hydraulic lash adjuster and the forces associated with the plunger with the check valve closed. A co-ordinate system (*yoz*) is fixed on the follower as shown in the figure. These forces are: spring force, F_{ps}, the inertia, I_p, oil pressure forces, F_{p1} and F_{po}, valve stem force, P, and frictions, F_{f1} and F_{f2}. The balance of these forces in the z direction gives

$$\sum F_z = -I_p - P + F_{ps} + F_{po} + F_{p1} - F_{f1} - F_{f2} = 0 \tag{3.1}$$

1. The spring force F_{ps} is caused by the precompression δ_p of the check valve spring and the plunger displacement z and can be written as

$$F_{ps} = k_{ps}\left(z - \delta_p\right) \tag{3.2}$$

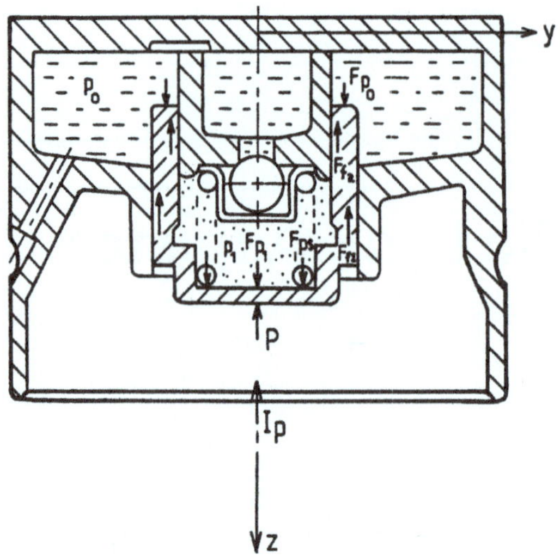

Fig. 3.1 Lash adjuster – check valve closed

where k_{ps} is the stiffness of the plunger spring.

2. The inertia of the plunger I_p is given by

$$I_p = m_p\left(\ddot{l}_f + \ddot{z}\right) \tag{3.3}$$

where m_p is the mass of the plunger.

3. F_{p1} and F_{po} are the forces due to oil pressure and are given by the product of the pressure and corresponding area as

$$F_{po} = A_{pr}\,p_o = \pi p_o\left[r_1^2 - (r_2 + h_2)^2\right]$$
$$F_{p1} = A p_1 = \pi p_1 (r_2 + h_2)^2$$

where p_o is the oil pressure in the follower reservoir and is equal to the inlet oil pressure. p_1 is the oil pressure in the plunger chamber and can be deduced from the conservation of the flow between the follower reservoir and the plunger chamber. The flow conservation means that the flow squeezed out from the clearance between the plunger and the inner cylinder equals the decreasing rate of the plunger chamber volume due to the movement of the plunger. Figure 3.2 is a sketch drawn for deducing the pressure p_1. The parameters adopted in the analysis are also presented. This leads to

$$p_1 = p_o - \frac{4\eta H_2}{h_2^3(2r_2 + h_2)}\left(3r_2^2 + 3r_2 h_2 + h_2^2\right)\dot{z} \tag{3.4}$$

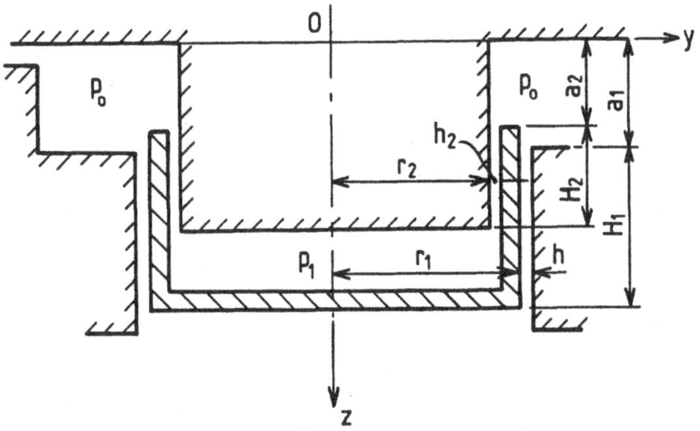

Fig. 3.2 Geometry adopted to calculate the pressure p_1

4. The force P applied by the valve stem arises from the spring force and the inertia. Thus

$$P=\left(\ddot{l}_f+\ddot{z}\right)\left(m_v+\frac{m_s}{3}\right)+k_{vs}\left(l_f+\delta+z\right)+F_v \tag{3.5}$$

where m_s is the valve spring mass, k_{vs} is the stiffness of the valve spring, δ is the precompression of the valve spring, F_v is the friction arising from the interface of the valve stem and the valve guide and is given by equation (7.7).

5. Frictions between the plunger/bush F_{f1} and the plunger/inner cylinder F_{f2} are assumed due to the shear of the oil flow between these surfaces. So

$$F_{f1}=2\pi\eta_o r_1 H_1\frac{\partial V_1}{\partial y}\Big|_{y=r_1}=2\pi\eta_o r_1 H_1\left(\frac{p_o h_1}{2\eta_o H_1}-\frac{\dot{z}}{h_1}\right) \tag{3.6}$$

$$F_{f2}=2\pi\eta_o r_2 H_2\frac{\partial V_2}{\partial y}\Big|_{y=r_2+h_2}=2\pi\eta_o H_2(r_2+h_2)\left(\frac{p_1-p_o}{2\eta_o H_2}h_2+\frac{\dot{z}}{h_2}\right) \tag{3.7}$$

3.2.2 *The governing equation of the displacement*

The governing equation associated with the displacement z, the velocity $\dot{z}$, and the acceleration z, of the plunger was deduced by substituting equation (3.1) with equations (3.2), (3.3), (3.4), (3.5), (3.6), and (3.7). This gives

$$\ddot{z}+C\dot{z}+Kz=B \tag{3.8}$$

where

$$B=\frac{1}{M}\left[\pi r_1(r_1-h_1)p_o+k_{ps}\delta_p-\pi\eta_o D_v L_v\frac{\dot{l}_f}{C_v}-k_{vs}\left(l_f+\delta\right)\right]-\ddot{l}_f$$

$$M=m_v+m_p+\frac{m_s}{3}$$

$$C=\frac{2\pi\eta_o}{M}\left[\frac{(r_2+h_2)H_2}{h_2}-\frac{r_1H_1}{h_1}+\frac{D_vL_v}{2C_v}+\frac{2r_2H_2(r_2+h_2)\left(3r_2^2+3r_2h_2+h_2^2\right)}{h_2^3(2r_2+h_2)}\right]$$

$$K=\frac{1}{M}\left(k_{vs}+k_{ps}\right)$$

Equation (3.8) is a second order ordinary differential equation associated with the plunger displacement with known initial values. As the follower and the valve remained steady on the cam base circle, it is quite reasonable to assume the initial displacement z and the velocity $\dot{z}$ of the plunger to be zero. As the cam lift l_c is provided in a discrete form, with a list of cam angles and corresponding lifts, this equation is solved by numerical analysis.

3.2.3 Numerical analysis

The numerical solution to equation (3.8) requires expanding $z(t+\delta t)$ and $z(t-\delta t)$ in the vicinity of time t in a Taylor series expansion. Thus the velocity and the acceleration of the plunger are expressed as

$$\begin{aligned}\ddot{z}_i&=d_c\left[\left(z_{i+1}-z_i\right)d_{i-1}^2+\left(z_i-z_{i-1}\right)d_i^2\right]\\ \dot{z}_i&=2d_c\left[\left(z_{i+1}-z_i\right)d_{i-1}+\left(z_i-z_{i-1}\right)d_i\right]\end{aligned}\tag{3.9}$$

where

$$d_i=t_{i+1}-t_i$$

$$d_{i-1}=t_i-t_{i-1}$$

$$d_c=\frac{1}{d_{i-1}d_i\left(d_i+d_{i-1}\right)}$$

By substituting $\dot{z}$ and $\ddot{z}$ given by equation (3.9), equation (3.8) is converted into the following system of linear equations with a series of unknowns z_j.

$$\left[A_{i,j}\right]\left\{z_j\right\}=\left\{B_i\right\}\qquad i,j=1,2,3,\ldots,m\tag{3.10}$$

where

$$\begin{aligned}A_{i,j}&=0 &&\text{if } j<i-1 \text{ or } j>i+1\\ A_{i,i-1}&=d_cd_i\left(2-Cd_i\right)\\ A_{i,i}&=Cd_c\left(d_i^2-d_{i-1}^2\right)+2d_c\left(d_{i-1}+d_i\right)+K\\ A_{i,i+1}&=d_cd_i\left(2+Cd_{i-1}\right)\\ B_i&=B\left(t_i\right)\end{aligned}$$

The coefficient matrix $[A_{i,j}]$ in the linear equation (3.10) is a tri-diagonal matrix, which can be solved by a Gaussian elimination approach.

3.3 Check valve open

The solution to equation (3.10) yields a series of displacements of the plunger in respect of each cam angle. For most of the cam event the plunger is compressed (or negative displacement according to the co-ordinate system in Fig. 3.1). This is, however, not true on the cam closing side when the cam lift is smaller than the plunger displacement. In this case, the follower tends to leave the cam and the load on the follower is released. In consequence the compressed plunger spring expands resulting in an increase of volume of the plunger chamber. This leads to a pressure drop in the chamber and consequently the check valve opens. Thus the oil flows into the chamber from the follower reservoir through the orifice between the chamber and the reservoir.

3.3.1 The dynamics of the plunger

When the check valve opens, the expansion of the plunger spring and the pressure of the oil pushes the plunger down against the valve stem. As the load applied by the plunger spring and oil pressure is much smaller than the valve spring and also due to gravity of the follower, the plunger remains steady on the end of the valve stem. In consequence the follower is pushed up to take up the clearance between the cam and the follower. In this case the movement of the follower depends upon the forces associated with both the plunger and the follower. A layout of the hydraulic lash adjuster and the forces associated with the plunger with the check valve opened are shown in Fig. 3.3. A co-ordinate system (*yoz*) was fixed on the follower as shown in the figure. These forces are: spring force F_{ps}, oil pressure forces F_{p1} and F_{po}, the force P due to the movement of the follower, and frictions F_{f1} and F_{f2}. The balance of these forces in the z direction yields

$$\sum F_z = -P + F_{ps} + F_{po} + F_{p1} - F_{f1} - F_{f2} = 0 \tag{3.11}$$

1. Spring force F_{ps} due to the precompression δ_p of the check valve spring and the plunger displacement z is given by

$$F_{ps} = k_{ps}\left(z - \delta_p\right) \tag{3.12}$$

2. Forces F_{p1} and F_{po} due to the oil pressure are the same as those when the check valve closes (see Section 3.2.1, list point 3). p_1 is the oil pressure in the plunger chamber and can be deduced from the conservation of the flow between the follower reservoir and the plunger chamber. The flow conservation means that the oil flow which flows into the plunger chamber through the orifice should equal the increasing rate of volume change of the chamber due to the movement of the plunger. Figure 3.4 is a sketch drawn for deducing the pressure p_1. The parameters adopted in the analysis are also presented. The conservation of the flow gives

$$p_1 = p_o - \frac{\pi^2 \rho}{2\xi^2 A_0^2}(r_2 + h_2)^4 \dot{z}^2 \tag{3.13}$$

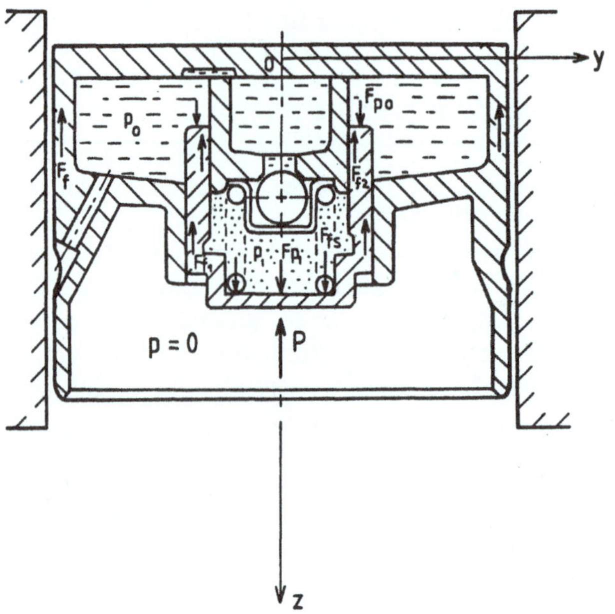

Fig. 3.3 Lash adjuster – check valve open

where A_o is the area of the orifice, ρ is the oil density and ξ is the flow loss coefficient (between 0.6 and 0.65).

3. The force P arises from the friction F_f between the follower and its bush and the inertia of the follower excluding the plunger. Thus

$$P = m_t \ddot{z} + F_f \tag{3.14}$$

where m_t is the mass of the follower excluding the plunger.

4. The friction between the follower and its bush F_f is caused by the shear of the oil flow in two sections of the follower B_1 and B_2 as shown in Fig. 3.4. Following the same procedure for deducing F_{f1} and F_{f2}, F_f is obtained as

$$F_f = \pi\eta_o D_f \left[B_1 \frac{\partial V_1}{\partial y}\bigg|_{y=\frac{D_f}{2}} + B_2 \frac{\partial V_2}{\partial y}\bigg|_{y=\frac{D_f}{2}} \right] = -\pi\eta_o D_f (B_1 + B_2)\frac{\dot{z}}{h} \tag{3.15}$$

5. Frictions between the plunger/bush F_{f1} and the plunger/inner cylinder F_{f2} are the same as those given by equations (3.6) and (3.7). Substituting (3.6) and (3.7) with (3.13) gives

$$F_{f1} + F_{f2} = \pi r_1 h_1 p_o + 2\pi\eta_o \left[\frac{(r_2 + h_2)H_2}{h_2} - \frac{r_1 H_1}{h_1} \right] \dot{z} - \frac{\pi^3 \rho h_2 (r_2 + h_2)^5}{2\xi^2 A_o^2 \dot{z}^2} \tag{3.16}$$

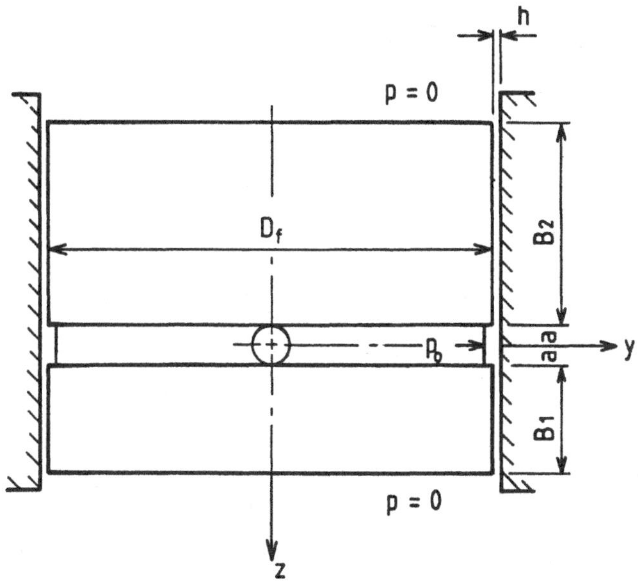

Fig. 3.4 Geometry for calculating follower/bush friction

3.3.2 The governing equation of the displacement

The governing equation associated with displacement z, velocity $\dot{z}$, and acceleration $\ddot{z}$ of the plunger is deduced by substituting equation (3.11) with equations (3.12), (3.14), (3.15), and (3.16). This leads to

$$\ddot{z}+C_1\dot{z}+C_2\dot{z}^2+K_p z=F \tag{3.17}$$

where

$$F=\frac{1}{m_t}\left[\pi r_1(r_1-h_1)p_o+k_{ps}\delta_p\right]$$

$$m_t=m_f-m_p$$

$$C_1=\frac{2\pi\eta_o}{m_t}\left[\frac{(r_2+h_2)H_2}{h_2}-\frac{r_1H_1}{h_1}+\frac{D_f(B_1+B_2)}{2h}\right]$$

$$C_2+\frac{1}{2\xi^2A_o^2m_t}\left[\pi^3\rho r_2(r_2+h_2)^5\right]$$

$$K_p=\frac{k_{ps}}{m_t}$$

Equation (3.17) is a non-linear second order ordinary differential equation associated with the plunger displacement with known initial values. This equation can only be solved by numerical integration.

3.3.3 Numerical analysis

Letting Y_1 equal z and Y_2 equal $\dot{z}$, equation (3.17) can be written in the following form.

$$\begin{aligned} \dot{Y}_1 &= Y_2 \\ \dot{Y}_2 &= F - C_1 Y_2 - C_2 Y_2^2 - K_p Y_1 \end{aligned} \tag{3.18}$$

With known initial values of z_i, $\dot{z}_i$ and $\ddot{z}_i$ at t_i, the plunger displacement z_{i+1} at the next time step t_{i+1} is evaluated from the following equation

$$z_{i+1} = z_i + \dot{z}_i \left(t_{i+1} - t_i\right) + \frac{\ddot{z}_i}{2}\left(t_{i+1} - t_i\right)^2 \tag{3.19}$$

Numerical integration of equation (3.18) is also performed with the sub-routine D02EAF of NAG – FORTRAN LIBRARY (Mark 14) to evaluate the plunger displacement z_{i+1} at the next time step t_{i+1}. The differences between the results of the numerical integration and the estimations by equation (3.19) are less than one per cent.

3.4 The numerical solution procedure

First, the solution is started on the cam base circle and carried out for the whole cam cycle with the assumption of the check valve closed [equation (3.10)]. At the point where the displacement of the plunger is greater than the cam lift the check valve opens. Then equation (3.18) is integrated. If the displacement of the plunger estimated by equation (3.18) exceeds the cam lift, it is assumed that the plunger keeps contact with the valve stem. Therefore, the cam lift is taken as the plunger displacement.

Chapter 4

The Maximum Hertzian Stresses

4.1 Introduction

In order to reduce the likelihood of scuffing and pitting failure of the cam and follower it is important to keep the contact stress as low as possible. The theory used to determine the contact stress was established by Hertz. The maximum Hertzian stress is adopted as a design criterion in cam and follower design. Although the stress distribution and the maximum stress associated with the contact of a taper cam and a domed follower might not be the same as predicted by the Hertzian theory [1], this theory can still provide good results for a quick estimation of the stress level.

In addition to this, elastohydrodynamic lubrication analysis has shown that the Hertzian pressure distribution is a very good approximation to the oil pressure distribution within the contact region of the cam and follower. It has also shown the friction of the junction to be dominated by the shear stresses within the contact region which can also be predicted from the Hertzian theory using the concept of a limiting friction coefficient.

4.2 The contact geometry

For the ZETEC engine a taper cam and domed follower is adopted. The contact between the cam and follower is equivalent to the contact of an elastic sphere with a segment of an elastic cone whose half taper is equal to the cam taper. A stress analysis based on this model was undertaken.

The full analysis of such a problem, however, requires a complex numerical analysis and a long central processing unit (CPU) time which might not be acceptable to cam designers. A reasonable simplification to the problem is made by assuming the contact between the cam and follower to be equivalent to an elastic sphere, with the radius of curvature of the domed follower R_f, in contact with an infinite long elastic cylinder with the radius R_c of curvature of the cam. Furthermore, Hertzian theory indicates that such a contact is geometrically equivalent to an ellipsoid, with principal radii of curvature, R_x and R_y, in contact with a plane. Figure 4.1 shows the contact ellipse predicted by the Hertzian theory; a is the semi-major axis of ellipse of the contact and b is the semi-minor axis of ellipse of the contact. According to the Hertzian theory [1]

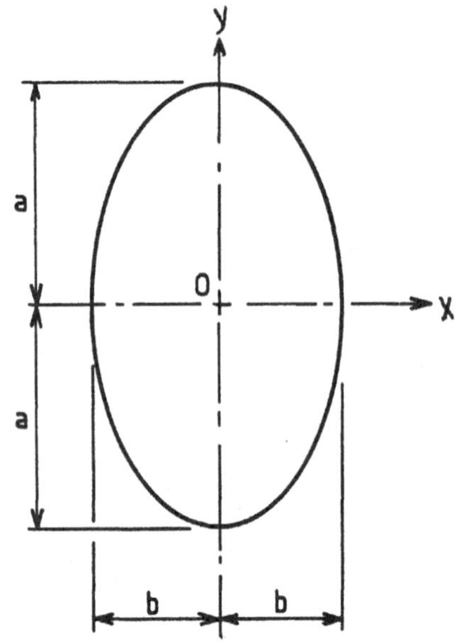

Fig. 4.1 Contact ellipse

$$a = \left(\frac{6k^2 \bar{\varepsilon} RW}{\pi E'} \right)^{\frac{1}{3}} \tag{4.1}$$

$$b = \left(\frac{6\bar{\varepsilon} RW}{\pi k E'} \right)^{\frac{1}{3}} \tag{4.2}$$

where k is the ellipticity ratio of the semi-major axis to semi-minor axis and a is greater than b.

$$k = \frac{a}{b} = \frac{2\bar{F} - (1+\Gamma)\bar{\varepsilon}}{\bar{\varepsilon}(1-\Gamma)} \tag{4.3}$$

$$\bar{\varepsilon} = \int_0^{\frac{\pi}{2}} \left[1 - \left(1 - \frac{1}{k^2} \right) sin^2\varphi \right]^{\frac{1}{2}} d\varphi \tag{4.4}$$

$$\bar{F} = \int_0^{\frac{\pi}{2}} \left[1 - \left(1 - \frac{1}{k^2} \right) sin^2\varphi \right]^{-\frac{1}{2}} d\varphi \tag{4.5}$$

$$\Gamma = \frac{R_y - R_x}{R_x + R_y} = \frac{1 - R_x / R_y}{1 + R_x / R_y} \tag{4.6}$$

$$R = \frac{R_x R_y}{R_x + R_y} \tag{4.7}$$

$$R_x = \frac{R_f R_c}{R_f + R_c} \tag{4.8}$$

$$R_y = R_f \tag{4.9}$$

4.3 The maximum Hertzian stress

The maximum Hertzian stress $p_{\max}$ is given by the expression

$$p_{\max} = \frac{E'}{2\pi}\left[\frac{3W}{E'R^2}\right]^{\frac{1}{3}} = \frac{3W}{2\pi ab} \tag{4.10}$$

where E', the equivalent elastic constant, is given by

$$E' = \frac{2E_c E_f}{\left(1 - v_c^2\right)E_f + \left(1 - v_f^2\right)E_c} \tag{4.11}$$

where
E_c is the Young's modulus of cam material;
E_f is the Young's modulus of follower material;
v_c is the Poisson's ratio of follower material;
v_f is the Poisson's ratio of follower material.

4.4 The pressure distribution

The pressure distribution p within the Hertzian contact region is semi-elliptical given by

$$p(x, y) = p_{\max}\left(1 - \frac{x^2}{b^2} - \frac{y^2}{a^2}\right)^{\frac{1}{2}} \tag{4.12}$$

where x and y are the co-ordinates from the centre of the contact.

4.5 The solution procedure

In the solution to the problem the key unknown is the ellipticity ratio which couples several equations. In the present analysis the iteration started with an initial value of the ellipticity ratio which was given by the following approximate formulae.

$$k = \left(\frac{R_y}{R_x} \right)^{\frac{2}{3}} \tag{4.13}$$

With the initial ellipticity ratio k, $\overline{\mathcal{E}}$, and $\overline{F}$ are estimated from numerical integration of equations (4.4) and (4.5). Then a new ellipticity ratio is calculated with equation (4.3). If the relative difference between the initial k and the new k exceeds the tolerance (0.1 per cent) the later one is taken as the initial k and this procedure is repeated until the value of ellipticity ratio k converged. With the latest values of k and $\overline{\mathcal{E}}$, a and b are calculated from equations (4.1) and (4.2).

4.6 References

1. **Hamrock, B. J.** and **Dowson, D.** (1981) *Ball Bearing Lubrication – the Elastohydrodynamics of Elliptical Contacts* (John Wiley & Son, New York).

Chapter 5

Asperity Interactions

5.1 Introduction

Both theoretical and experimental studies have shown that the surface roughness has significant influences on the tribological characteristics of the cam and follower. The asperity interaction model adopted in the present analysis is that developed by Greenwood and Tripp [1]. The asperity contact force (or the load carried by asperities) and the real area of contact can be predicted by this model.

5.2 The asperity contact force

The asperity interaction model is based on the assumption of a Gaussian distribution of asperity heights and contact radius of curvature of the symmetric asperities. In the present model only elastic asperity deformation is considered.

$$W_a = \frac{8\sqrt{2}}{15}\pi\left(\eta\beta\sigma\right)^2 E'\sqrt{\frac{\sigma}{\beta}}AF_{\frac{5}{2}}\left[\frac{h}{\sigma}\right] \tag{5.1}$$

where
η is the asperity density;
σ is the composite roughness height, $\sqrt{\sigma_c^2+\sigma_f^2}$;
β is the asperity radius of curvature;
E' is the equivalent elastic constant;
A is the apparent area of contact;
h is the nominal separation between two reference surfaces;
$F_{\frac{5}{2}}$ is the asperity contact function.

5.3 The real area of contact

The real area of contact A_a is related to the apparent area of contact A through the following relationship:

$$A_a = \pi^2\left(\eta\beta\sigma\right)^2 AF_2\left[\frac{h}{\sigma}\right] \tag{5.2}$$

where F_2 is another asperity contact function.

5.4 The asperity contact functions

The asperity contact functions for a Gaussian distribution of asperity height described in the above sections are defined by the following integrations:

$$F_2\left(\frac{h}{\sigma}\right)=\int_{\frac{h}{\sigma}}^{\infty}\left(s-\frac{h}{\sigma}\right)^2 e^{-\frac{s^2}{2}}ds \tag{5.3}$$

$$F_{\frac{5}{2}}\left(\frac{h}{\sigma}\right)=\int_{\frac{h}{\sigma}}^{\infty}\left(s-\frac{h}{\sigma}\right)^{\frac{5}{2}} e^{-\frac{s^2}{2}}ds \tag{5.4}$$

In order to reduce the CPU time for numerical integrations, curve-fitted formulae for the two functions are produced. The details have been recorded in reference [2].

Let H equal $\left(\frac{h}{\sigma}\right)$, we have

$$F_2(H)=\begin{cases} d_1 e^{d_2 \ln(H_1-H)+d_3[\ln(H_1-H)]^2} & \text{if} \quad H\le 2.0 \\ d_4 e^{d_5 \ln(H_2-H)+d_6[\ln(H_2-H)]^2} & \text{if} \quad 2.0<H\le 3.5 \\ d_7 (H_3-H)^{d_8} & \text{if} \quad 3.5<H\le 4.0 \\ 0 & \text{if} \quad H>4.0 \end{cases}$$

$$F_{\frac{5}{2}}(H)=\begin{cases} f_1 e^{f_2 \ln(H_1-H)+f_3[\ln(H_1-H)]^2} & \text{if} \quad H\le 2.0 \\ f_4 e^{f_5 \ln(H_2-H)+f_6[\ln(H_2-H)]^2} & \text{if} \quad 2.0<H\le 3.5 \\ f_7 (H_3-H)^{f_8} & \text{if} \quad 3.5<H\le 4.0 \\ 0 & \text{if} \quad H>4.0 \end{cases}$$

Where $H_1 = 9.0$, $H_2 = 8.0$, and $H_3 = 4.0$

$d_1 = 0.14704 \times 10^{-40}$ $f_1 = 0.11755 \times 10^{-39}$
$d_2 = 0.70373 \times 10^{2}$ $f_2 = 0.67331 \times 10^{2}$
$d_3 = -0.12696 \times 10^{2}$ $f_3 = -0.11699 \times 10^{2}$
$d_4 = 0.74104 \times 10^{-21}$ $f_4 = 0.15827 \times 10^{-20}$
$d_5 = 0.30813 \times 10^{2}$ $f_5 = 0.29156 \times 10^{2}$
$d_6 = -0.36470 \times 10^{1}$ $f_6 = -0.29786 \times 10^{1}$
$d_7 = 0.88123 \times 10^{-4}$ $f_7 = 0.11201 \times 10^{-3}$
$d_8 = 0.21523 \times 10^{1}$ $f_8 = 0.19447 \times 10^{1}$

5.5 The solution procedure

The solution procedure for the asperity interaction force and real contact area is straightforward. For a given nominal separation h the force and the area can be predicted by equations (5.1) and (5.2) respectively. In the present analysis the central film thickness h_{cen} given by equation (6.9) is taken to be the nominal separation between the cam and follower. The area of the ellipse of the contact is taken to be the apparent contact area.

5.6 References

1. **Greenwood, J. A.** and **Tripp, J. H.** (1970–1971) The contact of two nominally flat rough surfaces. *Proc. Instn Mech. Engrs,* **185**, 625–633.
2. **Zhu, G.** (1988) A theoretical and experimental study of the tribology of a cam and follower. PhD thesis, Mechanical Engineering Department, Leeds University, UK, May.

Chapter 6

The Oil Film Thickness

6.1 Introduction

Predicting the oil film thickness between a taper cam and domed follower is one of the major tasks for the tribological analysis. The analysis of such a problem requires a full numerical elastohydrodynamic lubrication study of a non-steady state elliptical contact with spin caused by follower rotation. Such an analysis, however, will involve tremendous computing time and the possibility of numerical difficulties. Obviously this technique is only suitable for research oriented purposes. Fortunately a robust technique for estimating the oil film thickness associated with elliptical contacts with the direction of the lubricant entrainment at some intermediate angle to the major axis of the Hertzian contact ellipse has been developed by Chittenden *et al.* [1, 2]. This makes it possible to perform a quick quasi-static analysis for predicting oil film thicknesses between a taper cam and domed follower.

6.2 The lubricant entraining velocity

As introduced in Chapter 5 the contact between a taper cam and a domed follower equates to an elastic sphere, with the radius of curvature of the domed follower R_f, in contact with an infinite long elastic cylinder with the radius of curvature R_c of the cam. Furthermore, such a contact is geometrically equivalent to an ellipsoid, with principal radii of curvature R_x and R_y, in contact with a plan [3].

The lubricant entraining velocity is one of the major parameters for predicting the oil film thickness. Due to the rotation of the follower, however, the entraining velocities vary in both magnitude and direction at each point of the contact area. In the present analysis the average entraining velocity, which is the velocity at the centre of the contact ellipse, is taken to be the lubricant entraining velocity for estimating the central and the minimum film thickness. Figure 6.1 shows an elliptical elastohydrodynamic lubrication conjunction. If the cam and the follower have surface velocity components u_c,u_f and v_c, v_f in the directions of the principal axes of the Hertzian ellipse (x,y), the entraining velocities in these directions are

$$u = \frac{1}{2}\left(u_c + u_f\right) \tag{6.1}$$

$$v = \frac{1}{2}(v_c + v_f) \tag{6.2}$$

For a general entraining direction e, inclined at an angle θ to the minor axis x of the contact, as shown in Fig. 6.1,

$$\theta = \arctan\left(\frac{v}{u}\right) \tag{6.3}$$

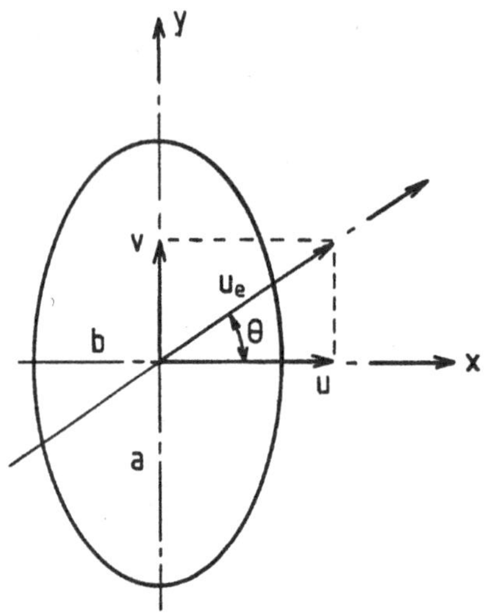

Fig. 6.1 Contact area with lubricant entrainment at an angle to the principal area

The effective entraining velocity u_e is then given by

$$u_e = \sqrt{u^2 + v^2} \tag{6.4}$$

According to the kinematic analysis in Chapter 2 the entraining velocity components are given by

$$u_c = V_c \tag{6.5}$$

$$v_c = 0 \tag{6.6}$$

$$u_f = V_{fm} + \Omega(\delta_t + y) \tag{6.7}$$

$$v_f = \Omega(e - x) \tag{6.8}$$

where V_c, V_{fm}, and δ_t are given by equations (2.1), (2.2) and (2.5) respectively.

6.3 The central and the minimum film thickness

The theoretical analysis of the isothermal elastohydrodynamic lubrication of concentrated contact with lubricant entrainment at some intermediate angle to the minor axis of the Hertzian ellipse shows that the central film thickness h_{cen} can be expressed as

$$h_{\text{cen}} = 4.31\left\{1-\exp\left[-1.23\left(\frac{R_s}{R_e}\right)^{\frac{2}{3}}\right]\right\}U_e^{0.68}G^{0.49}W_e^{-0.073}R_e \tag{6.9}$$

and the minimum film thickness h_{min} is given by

$$h_{\text{min}} = 3.68\left\{1-\exp\left[-0.67\left(\frac{R_s}{R_e}\right)^{\frac{2}{3}}\right]\right\}U_e^{0.68}G^{0.49}W_e^{-0.073}R_e \tag{6.10}$$

where

G is the dimensionless materials parameter, $\alpha E'$;
E' is the equivalent elastic constant;
α is the viscosity–pressure coefficient, $\eta = \eta_o e^{\alpha p}$;
η_o is the dynamic viscosity at atmospheric pressure;
W_e is the dimensionless load parameter, $W_h / \left(E'R_e^2\right)$;
W_h is the hydrodynamic load, $W - W_a$;
W_a is the asperity load;
U_e is the dimensionless speed parameter, $\eta_o u_e / \left(E'R_e\right)$;
R_e is the effective radius of curvature in the direction of lubricant entrainment and is given by

$$R_e = \frac{R_x R_y}{R_x \sin^2\theta + R_y \cos^2\theta} \tag{6.11}$$

R_s is the effective radius of curvature in the direction of side-leakage and is given by

$$R_s = \frac{R_x R_y}{R_x \cos^2\theta + R_y \sin^2\theta} \tag{6.12}$$

R_x, R_y are equivalent radii of curvature in x, y directions and are given by equations (4.8) and (4.9) respectively.

6.4 The solution procedure

As the hydrodynamic load is unknown, the film thicknesses are predicted within an iterative loop. The cam load is taken to be an initial value of the hydrodynamic load to estimate an initial central film thickness. With the assumption of the nominal separation between two reference surfaces being equal to the central film thickness the asperity load is evaluated by equation (5.1). Then a new central film thickness is re-evaluated with the new hydrodynamic load W_h, which is known from the difference between the cam load W and the asperity load W_a. This procedure is repeated until the film thickness calculated between two iterations is convergent (with a 0.1 per cent tolerance).

6.5 References

1. **Chittenden, R. J., Dowson, D., Dunn, J. F.,** and **Taylor, C. M.** (1985) A theoretical analysis of the isothermal elastohydrodynamic lubrication of concentrated contacts: I. Direction of lubricant entrainment coincident with the major axis of the Hertzian ellipse. *Proc. R. Soc. Lond.*, **A397**, 245–269.

2 **Chittenden, R. J., Dowson, D., Dunn, J. F.,** and **Taylor, C. M.** (1985) A theoretical analysis of the isothermal elastohydrodynamic lubrication of concentrated contacts: II. General case, with lubricant entrainment along either principle axis of the Hertzian ellipse or at some intermediate angle. *Proc. R. Soc. Lond.*, **A397**, 271–294.

3. **Dowson, D.** and **Higginson, G. R.** (1966) *Elastohydrodynamic Lubrication* (Pergamon Press).

Chapter 7

Friction and Power Loss of the Valve Train

7.1 Introduction

Estimating the power loss of the valve train of the ZETEC engine is one of the essential functions of the present theoretical model. The power loss associated with the valve train is due to the frictions arising from the four interfaces between cam/follower, follower/bush, valve stem/guide and valve stem/seal. The friction model for each interface will be discussed in the following sections.

7.2 The friction of the cam and follower interface

The friction of the junction between the cam and follower is caused by two different mechanisms. One is due to the shear of the lubricant and the other is due to the asperity contact. In the present model these two parts of the friction are estimated separately and then summed with regard to the proportion of the asperity contact area.

7.2.1 The friction due to shear of the lubricant

Elastohydrodynamic lubrication theory has shown that the friction due to the shear of the lubricant within the conjunction of the cam and follower is dominated by the sliding contribution from the 'Hertzian' region which here is assumed to be parallel and subjected to the Hertz pressure distribution. The central film thickness is taken to be the separation of the two parallel surfaces.

The rheological model developed by Evans and Johnson [1] is adopted in the present analysis in which non-Newtonian effects of the lubricant are taken into consideration. So the shear stress between the two parallel surfaces is given by

$$\tau = \begin{cases} \eta \dfrac{V_s}{h_{\text{cen}}} & \text{if} \quad \tau \le \tau_o \\ \tau_o + \gamma p & \text{if} \quad \tau_o < \tau \le \tau_L \\ \tau_L & \text{if} \quad \tau > \tau_L \end{cases} \tag{7.1}$$

where
τ_o is the Eyring stress and varies with temperature and pressure;
τ_L is the limiting shear stress depending upon the mean pressure;
γ is the rate of change of shear stress with pressure;
V_s is the sliding velocity between the cam and follower ($u_c - u_f$);
p is the local pressure;
α is the viscosity–pressure coefficient;
η_o is the dynamic viscosity at atmospheric pressure;
η is the dynamic viscosity of the lubricant. The Barus relationship for the viscosity and pressure is given by

$$\eta = \eta_o e^{\alpha p} \tag{7.2}$$

Thus the integration of the shear stresses over the whole Hertzian region yields the friction due to the shear of the lubricant within the conjunction of the cam and follower. That is

$$F_c^s = \int_{-a}^{a} \int_{-b}^{b} \tau \mathrm{d}x\mathrm{d}y \tag{7.3}$$

7.2.2 *The friction due to asperity contact*

The friction arising from the asperity contact is assumed to be due to the asperity sliding under a boundary lubrication condition [2]. Thus the friction F_c^a can be expressed as

$$F_c^a = \mu_a W_a \tag{7.4}$$

where
μ_a is asperity sliding coefficient of friction;
W_a is the asperity load given by equation (5.1).

The summation of the frictions due to the shear of lubricant and the asperity contact with regard to the asperity contact area gives the friction F_c of the conjunction of the cam and follower as

$$F_c = F_c^a + \left(1 - \frac{A_a}{A}\right) F_c^s \tag{7.5}$$

where A_a is the real area of contact given by equation (5.2).

7.3 The friction of the follower and bush interface

The previous experimental study of a cam and bucket follower [3] indicated that the follower tilted as it moved up and down. Consequently, a theoretical model [3] for predicting the friction arising from the interface of the tappet and bore was developed with the assumption of a tilting follower under boundary lubrication conditions.

The current experimental evidence (see Section 16.3) also suggested that the tappet tilted as it travelled up and down in the tappet bore. The lubrication condition of the interface, however, was improved due to the two following factors: first, as a hydraulic lash adjuster was adopted in the valve train of the ZETEC engine, pressurized oil was fed into the gap between the follower and the bore; second, the tappet was designed to be free to rotate. Both factors were beneficial to the hydrodynamic lubrication. Thus the lubrication condition of the interface might be influenced by hydrodynamic, mixed, or boundary lubrication. The friction arising from the interface should therefore be evaluated according to its lubrication condition.

A full lubrication analysis of the tappet/bore interface involves complex numerical analysis and iterations and subsequently a large amount of computing time. This is obviously not suitable for the model to be developed for the current design purposes and is also far beyond the scope of the current research. Thus an equivalent model which is based upon the short bearing theory has been developed.

7.3.1 The tilting of the follower

Figure 7.1 shows a view of the central plane of a tilting tappet contacts with a cam. The forces W_{b1} and W_{b2} supporting the tappet can be calculated by

$$W_{b1} = W_{b2} - F_c$$

$$W_{b2} = \frac{d}{L_f} W$$

where
L_f is the follower length;
W is the load on the cam;
d is the distance from the contact point to the follower centre, $\sqrt{e^2 + \delta_t^2}$
F_c is the friction arising from the cam/follower interface.

The tappet/bore interface is considered as a bearing with a shaft (tappet) rotating and sliding up and down in a bush (tappet bore). In order to balance the reaction force W_{b2} the tappet will shift towards the right-hand side with an equivalent eccentricity e_2. Due to the tilting of the tappet this eccentricity is considered to be located in the centre of the top half of the tappet. In the same way, the tappet must move to the left-hand side with an equivalent eccentricity e_1 located in the centre of the bottom half of the tappet. Because of the tilting of the tappet the hydrodynamic effect arising from the translation of the tappet due to the convergent wedge as shown in Fig. 7.1 will resist the tilting of the tappet. This is equivalent to reducing the eccentricities e_1 and e_2. Obviously a full lubrication analysis of the tappet/bore interface involves a complex numerical analysis and the determination of the lubrication boundary condition, which is not known if the Reynolds boundary condition is still valid in this situation, and therefore is not suitable for the model to be developed for design purposes. In order to take into account this

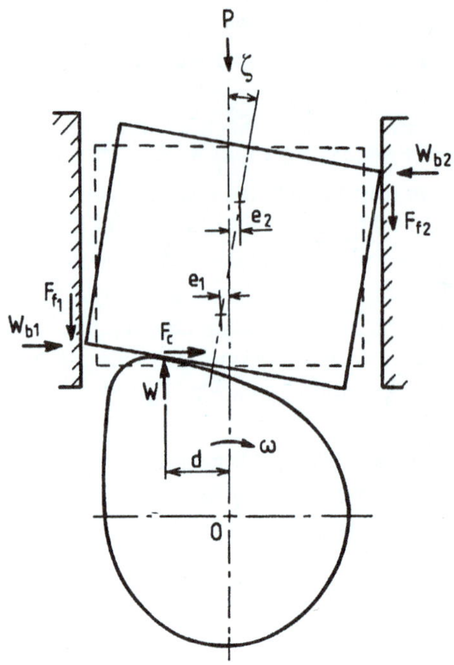

Fig. 7.1 The contact of a tilting follower and a cam

effect, the equivalent surface velocity has been adopted in the current model which is a resultant surface velocity due to the rotation and translation of the tappet. The hydrostatic effect of the inlet oil hole on the lubrication of the tappet/bore interface is neglected in the current model. e_1 and e_2 can be estimated from short bearing theory [4] given by the following equations:

$$W_{b1} = \frac{\eta_b U L_f^3}{4C_f^2} \frac{\varepsilon_1}{\left(1-\varepsilon_1^2\right)^2} \sqrt{\pi^2\left(1-\varepsilon_1^2\right)^2 + 16\varepsilon_1^2}$$

$$W_{b2} = \frac{\eta_b U L_f^3}{4C_f^2} \frac{\varepsilon_2}{\left(1-\varepsilon_2^2\right)^2} \sqrt{\pi^2\left(1-\varepsilon_2^2\right)^2 + 16\varepsilon_2^2}$$

where

ε_1 and ε_2 are dimensionless eccentricities (e_1/C_f) and (e_2/C_f).

U is the equivalent surface velocity, $\sqrt{\left(D_f \Omega/2\right)^2 + \dot{l}_f^2}$;

$\dot{l}_f$ is the translating velocity of the tappet, dl_f/dt ;

C_f is the radial clearance between the tappet and bore $\left(D_b - D_f\right)/2$;

D_f is the diameter of the follower;
D_b is the diameter of the tappet bore;
η_b is the lubricant viscosity within the tappet/bore gap;
Ω is the tappet rotational frequency.

So the tilting angle ζ of the tappet is given by

$$\zeta = \arctan\left[\frac{2(e_1+e_2)}{L_f}\right]$$

7.3.2 The friction model

From Fig. 7.1 it can be seen that the most critical contact locations are in the vicinity of the top right corner and bottom left corner due to the tilting of the follower. The minimum separations C_{b1} and C_{b2} between the tappet and the bore at the two positions can be estimated respectively from the following equations:

$$C_{b1} = C_f - e_1 - \frac{L_f}{4} tg\zeta$$

$$C_{b2} = C_f - e_2 - \frac{L_f}{4} tg\zeta$$

The friction arising from the interface is estimated separately from the top F_{f2} and the bottom F_{f1} half of the tappet. So the friction arising from the interface is given by

$$F_f = F_{f1} + F_{f2}$$

The ratio λ_{bf_i} of the separation C_{bi} to the composite surface roughness height σ_{fb} is adopted to identify the lubrication condition for each half of the tappet.

If λ_{bf_i} is greater than three, hydrodynamic lubrication dominates the lubrication of the interface. The friction of the interface of the tappet and bore arising from each half of the tappet can be evaluated from

$$F_{f1} = \left(\frac{\pi\eta_b D_f L_f U}{2C_f\sqrt{1-\varepsilon_1^2}}\right)\frac{i_f}{U}$$

$$F_{f2} = \left(\frac{\pi\eta_b D_f L_f U}{2C_f\sqrt{1-\varepsilon_2^2}}\right)\frac{i_f}{U}$$

If λ_{bf_i} is less than three but greater than zero, mixed lubrication is expected within the tappet/bore interface. In this case the friction consists of two parts: the shear of the lubricant and the contact of the surface. The percentage of each part contributing to the friction is dependent upon the rate of the overlap which is defined as follows:

$$\kappa_i = \frac{\lambda_{bf_i}}{3.0}$$

Thus the friction is given by

$$F_{f1} = \left[(1.0-\kappa_1)\mu_b W_{b1} + \kappa_1 \frac{\pi\eta_b D_f L_f U}{2C_f\sqrt{1-\varepsilon_1^2}}\right]\frac{i_f}{U}$$

$$F_{f2} = \left[(1.0-\kappa_2)\mu_b W_{b2} + \kappa_2 \frac{\pi\eta_b D_f L_f U}{2C_f\sqrt{1-\varepsilon_2^2}}\right]\frac{i_f}{U}$$

where μ_b is the friction coefficient of boundary lubrication.

If λ_{bf_i} is less than zero, the friction of the interface is dominated by the boundary lubrication. Then the friction is evaluated by

$$F_{f1} = \mu_b W_{b1}\frac{i_f}{U}$$

$$F_{f2} = \mu_b W_{b2}\frac{i_f}{U}$$

To summarize, the friction arising from the interface of the tappet and bore can be written as

$$F_{fi} = \begin{cases} \left[\dfrac{\pi\eta_b D_f L_f U}{2C_f\sqrt{1-\varepsilon_i^2}}\right]\dfrac{i_f}{U} & \text{if } \lambda_{bf_i} > 3.0 \\ \left[(1-\kappa_i)\mu_b W_{bi} + \kappa_i \dfrac{\pi\eta_b D_f L_f U}{2C_f\sqrt{1-\varepsilon_i^2}}\right]\dfrac{i_f}{U} & \text{if } 0.0 \le \lambda_{bf_i} < 3.0 \\ \mu_b W_{bi}\dfrac{i_f}{U} & \text{if } \lambda_{bf_i} < 0.0 \end{cases} \quad (7.6)$$

7.4 The friction of the valve and guide interface

The friction F_v arising from the interface of the valve stem and the valve guide is assumed to be due to the Couette flow of the lubricant within the gap and can be expressed as

$$F_v = \pi\eta_o D_v L_v \frac{\dot{l}_f + \dot{z}}{C_v} \tag{7.7}$$

where
L_v is the length of the valve guide;
D_v is the diameter of the valve stem follower;
C_v is the radial clearance between the valve stem and guide;
$\dot{z}$ is the velocity of the plunger relative to the follower.

7.5 The friction of the valve stem and valve seal interface

The model for predicting the friction F_s arising from the valve stem/seal interface will be developed later.

7.6 Power loss of the valve train

The power loss of the valve train is given by the product of the friction torque and the camshaft rotational frequency. The friction torque of the valve train is composed of two components. The first is the friction torque T_f, which is associated with the frictions of the reciprocating components and the eccentricity of the contact point to the camshaft centre. These frictions resist the movement of the follower and the valve. Consequently extra torque is required to drive the camshaft. As the reciprocating frictions are symmetrical to the central axis of the follower and the valve, they can be converted into concentrated frictions, F_f, F_v, and F_s, acting along the central axis as shown in Fig. 2.2. Thus the torque T_f is given by

$$T_f = \left(F_f + F_v + F_s\right)\left(R_b + l_c + \sqrt{R_f^2 - \delta_t^2}\right) tg\alpha \tag{7.8}$$

where
l_c is the cam lift;
R_b is the radius of the cam base circle;
α is given by equation (2.6);
δ_t is given by equation (2.5).

The second component is the rotational friction torque T_c which is due to the friction F_c of the junction of the cam and follower. From Fig. 2.2 this torque can be expressed as

$$T_c = F_c\left[(R_b + l_c)\cos\alpha - (1 - \cos\alpha)\sqrt{R_f^2 - \delta_t^2}\right] \tag{7.9}$$

Thus the total instantaneous friction torque in respect of the cam angle ϕ is given by

$$T_{\text{fric}} = T_f + T_c \tag{7.10}$$

The instantaneous power loss due to friction is given by the product of the friction torque T_{fric} and the camshaft rotational frequency ω. This can be integrated over a cam cycle to yield an average power loss H of the valve train.

$$H = \frac{1}{2\pi}\int_0^{2\pi} \omega T_{\text{fric}} d\phi \tag{7.11}$$

7.7 The solution procedure

The solution procedure for the frictions and power loss of the valve train is straightforward. Numerical integration is adopted for predicting the average power loss.

7.8 References

1. **Evans, C. R.** and **Johnson, K. L.** (1986) The rheological properties of elastohydrodynamic lubricant. *Proc. Instn Mech. Engrs, Part C,* **200** (5), 625–633.
2. **Nemlekar, P. R.** and **Cheng, H. S.** (1980) Traction in rough elastohydrodynamic contacts. In *Surface Roughness Effects in Hydrodynamic and Mixed Lubrication* (Eds M. S. Rohde and H. S. Cheng).
3. **Zhu, G.** (1988) A theoretical and experimental study of the tribology of a cam and follower. PhD thesis, Mechanical Engineering Department, Leeds University, UK, May.
4. **Pinkus, O.** and **Sternlicht, B.** (1961) *Theory of Hydrodynamic Lubrication* (McGraw-Hill Book Company).

Chapter 8

The Rotation of the Follower

8.1 Introduction

It is known that a rotating follower can reduce scuffing failure which is potentially the most serious form of damage. For the valve train of the ZETEC engine this is achieved by offsetting a taper cam to a domed follower. As described in the previous chapters, the rotation of the follower introduces an extra entraining velocity component which alters both the magnitude and direction of the lubricant entrainment relative to the point of contact of the cam and follower. In consequence the lubrication condition of the cam/follower interface is affected by this rotation. Predicting the rotational frequency of the follower over the whole cam cycle, therefore, is extremely important to the tribological analysis of the valve train of the ZETEC engine. In this chapter a theoretical model is developed to predict the follower rotational frequency.

8.2 The governing equation of the rotation of the follower

The fundamental principle of the rotation of the follower is that the cam and follower pair operates by traction transmission. A fraction of the power for driving the camshaft is transmitted to the follower due to the friction of the cam/follower interface resulting in the rotation of the follower. The frictions arising from the follower/bush and the valve stem/plunger interfaces resist the rotation of the follower. In addition, the acceleration or deceleration of the follower also affect the balance of the system due to the moment of inertia of the follower. Thus the governing equation of the follower rotation is deduced from the balance of these torques shown in Fig. 8.1. This is written as

$$J\dot{\Omega} = T_{fc} - T_{fb} - T_{fp} \tag{8.1}$$

where
J is the moment of inertia of the follower;
$\dot{\Omega}$ is the angular acceleration of the follower, $\mathrm{d}\Omega/(\mathrm{d}t)$;
T_{fc} is the driving torque due to the friction of the cam/follower interface;
T_{fb} is the resisting torque due to the friction of the follower/bush interface;
T_{fp} is the resisting torque due to the friction of the valve stem/plunger interface.

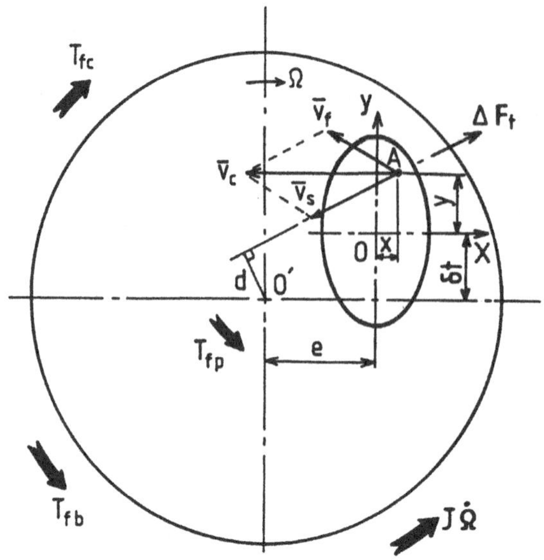

Fig. 8.1 The torques acting on the follower and in the contact ellipse

8.2.1 The driving torque of the cam/follower interface

The driving torque arises from the friction caused by the shear of the lubricant and the asperity contact within the conjunction of the cam and follower. The model developed for predicting the friction of the cam/follower interface has been described previously (see Chapter 7). This means that the present analysis is performed within the 'Hertzian' region which is assumed to be parallel and subjected to the Hertz pressure distribution. The central film thickness is taken to be the separation of the two parallel surfaces.

Figure 8.1 shows a contact ellipse with the centre of the contact O, having an eccentricity e, and shift δ_f, from the centre O' of the rotation of the follower. A co-ordinate system (xoy) is adopted with the origin being fixed at the centre of the contact ellipse. Due to the rotation of the follower, the sliding velocity $\bar{V}_s$ at each contact point is different in both magnitude and direction. Thus the contribution of each contact point to the driving torque is different. For a given point A, having the co-ordinate (x,y), the 'local' friction ΔF_t acts in a direction opposite to the sliding velocity. As illustrated in the figure, the sliding velocity is equal to the cam velocity vector $\bar{V}_c$ minus the follower velocity vector V_f. It should be pointed out that the velocity vectors shown in the figure are relative to the follower surface and therefore are in opposite directions to those shown in Fig. 2.1 in Chapter 2. The 'local' driving torque at point A, is given by

$$\Delta T_{fc} = d\Delta F_t \tag{8.2}$$

where d is the distance from the rotational centre O', of the follower to the vector ΔF_t and is written as

$$d = \frac{(V_c - V_{fm})(y+\delta_t) - \Omega\left[(e-x)^2 + (y+\delta_t)^2\right]}{V_s} \tag{8.3}$$

V_c and V_{fm} are given in Chapter 2 by equations (2.1) and (2.2) respectively. The sliding velocity V_s can be expressed as

$$V_s = \{[V_c - V_{fm} - \Omega(\delta_t + y)]^2 + \Omega^2(e-x)^2\}^{\frac{1}{2}} \tag{8.4}$$

ΔF_t represents the 'local' friction due to the shear of the lubricant and the asperity contact and is given by

$$\Delta F_t = \mu_a \frac{W_a}{A} + \left(1 - \frac{A_a}{A}\right)\tau \tag{8.5}$$

where W_a, A_a, and τ are given by equations (5.1), (5.2), and (7.1) respectively. The integration of ΔT_{fc} over the contact ellipse gives the driving torque T_{fc} as

$$T_{fc} = \int_{-a}^{a}\int_{-b}^{b} d\Delta F_t \mathrm{d}x\mathrm{d}y \tag{8.6}$$

Numerical integration is adopted to evaluate the driving torque given by equation (8.6). In the present analysis the contact ellipse is divided into a $m \times n$ mesh having $\Delta x = 2b/m$ width and $\Delta y = 2a/n$ length. The contribution to the driving torque arising from each small rectangular area can be represented by the product of the area $\Delta x \times \Delta y$ and an average value of four torques ΔT_{fc} which correspond to the four nodes (i, j), $(i+1, j)$, $(i, j+1)$, and $(i+1, j+1)$ of the rectangular A, as illustrated in Fig. 8.2. If the node (i, j) is outside the contact ellipse, the value of $(\Delta T_{fc})_{i,j}$ is taken to be zero. The sum of each individual contribution yields the driving torque as

$$T_{fc} = \sum_{i=1}^{m}\sum_{j=1}^{n}(\Delta T_{fc})_{i,j}\,\Delta x \Delta y \tag{8.7}$$

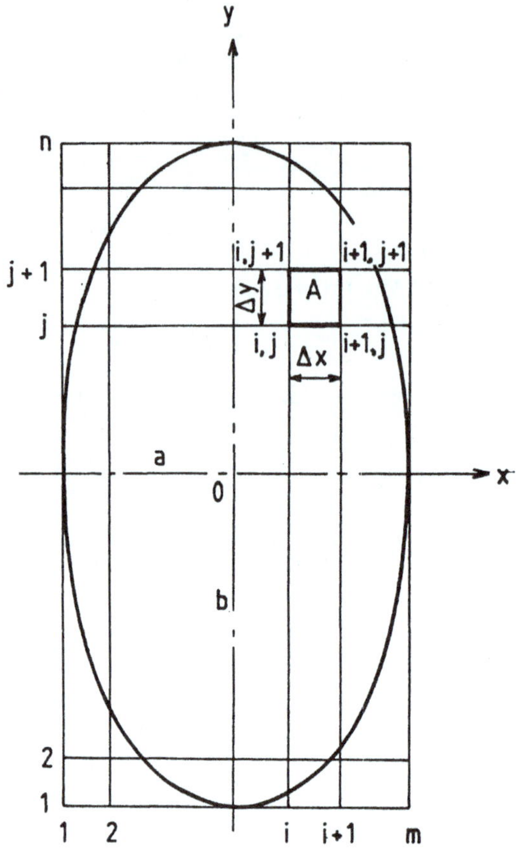

Fig. 8.2 The mesh structure of the contact ellipse for the numerical integration

8.2.2 The resisting torque of the follower/bush interface

The friction model of the follower/bore interface has been developed in Section 7.3. The friction against the rotation of the follower is the tangential component of the total friction of the interface. Therefore the friction can be written as

$$F_{fb} = \frac{\Omega D_f \left(F_{f1} + F_{f2}\right)}{2\dot{l}_f}$$

The product of the friction and the moment arm gives the resisting torque as

$$T_{fb} = \frac{\Omega D_f^2 \left(F_{f1} + F_{f2}\right)}{4\dot{l}_f} \tag{8.8}$$

where
U is the equivalent surface velocity (see Section 7.3.1);
F_{f1} and F_{f2} are evaluated from equation (7.6);

D_f is the diameter of the follower;
Ω is the rotational frequency of the tappet.

8.2.3 The resisting torque of the valve stem/plunger interface

As the shape of the plunger end is spherical and the end of the valve stem is flat, the contact between the two surfaces is equivalent to the contact of a ball with a plane. It is known that a pure spinning ball on a plane can hardly generate a hydrodynamic effect. A simple friction model therefore, is most appropriate for predicting the friction arising from the conjunction of the plunger and the valve stem end. According to Hertzian theory [1] the pressure distribution within the contact circle takes the form of

$$p = p_o \left[1 - \left(\frac{r}{a}\right)^2\right]^{\frac{1}{2}} \tag{8.9}$$

The maximum Hertzian stress p_o is given by

$$p_o = \frac{3P}{2\pi a^2} \tag{8.10}$$

and the radius of the circle a is given by

$$a = \left(\frac{3PR_p}{2E'}\right)^{\frac{1}{3}} \tag{8.11}$$

where R_p is the radius of curvature of the plunger end. The resisting torque T_{fp} due to the friction is given by the integration of the product of the friction and the distance to the centre of rotation.

$$T_{fp} = 2\pi\mu_c \int_o^a r^2 p_o \left[1 - \left(\frac{r}{a}\right)^2\right]^{\frac{1}{2}} \mathrm{d}r = \frac{3\pi\mu_c aP}{16} \tag{8.12}$$

where μ_c is the coefficient of friction. In the present analysis the limiting coefficient of friction of 0.08 is adopted.

8.3 The solution procedure

From equations (8.4) and (8.5) it is known that T_{fc} is an implicit function of the follower rotational frequency Ω. Therefore, an analytical solution to equation (8.1) is not available. With known initial values of Ω_i and $\dot{\Omega}_i$ at t_i, however, the follower rotational frequency Ω_{i+1} at the next time step t_{i+1} can be estimated from

$$\Omega_{i+1} = \Omega_i + \dot{\Omega}_i (t_{i+1} - t_i) \tag{8.13}$$

In addition the numerical solution of equation (8.1) is also performed using the sub-routine D02EAF of NAG – FORTRAN LIBRARY (Mark 14) to evaluate the follower rotational frequency Ω_{i+1}. The differences between the results of the numerical solution and the results of equation (8.13) were less than one per cent.

In order that the initial values of the follower rotational frequency and rotational acceleration can be determined, an iterative loop is introduced between two cam cycles. In the present analysis the solution procedure starts on the cam base circle, 90 degrees cam angle ahead the cam nose. An initial value Ω_o is estimated from equation (8.1) by assuming the angular acceleration $\dot{\Omega}_o$ to be zero. Along with the tribological analysis throughout the whole cam cycle the rotational frequency of the follower is also calculated at each time step from equation (8.13). Subsequently the follower rotational frequency at the starting point can be marched out by taking initial values to be those of the last point of the cam cycle. Thus, the analysis continues in a new cam cycle. In this cycle, at each time step the current rotational frequency is compared with the one in the previous cycle. Such analysis continues until the relative difference between the present rotational frequency and the one in the previous cycle is less than the tolerance (0.5 per cent).

8.4 References

1. **Johnson, K. L.** (1985) *Contact Mechanics* (Cambridge University Press).

Chapter 9

The Overall Solution Procedure and Input and Output Data

9.1 Introduction

The theoretical model which has been introduced in the previous chapters is coded in FORTRAN. In this section the overall solution procedure is described in a sequence as illustrated in the flow chart of the main program, which is shown in Fig. 9.1. The solution procedure of the main sub-routine in which the tribological analysis of the valve is performed is also detailed in this section. All the variables required in each solution step have been given as the input data or estimated in previous steps. The input data required to carry out the tribological analysis of the valve train and the output data are also detailed in this section. Equations are identified by two numbers which are separated by a dot. The number before the dot indicates the chapter in which the equation is presented. The other represents the sequence number of the equation.

9.2 Input data

The information required to carry out the tribological analysis on a specific valve train is identified by the input data. In this chapter the input data are categorized according to the component of the valve train. For each input parameter its nomenclature and the unit adopted in the present analysis are listed in Table 9.1. A dimensionless parameter is represented by a bar.

9.3 The kinematic analysis

With the cam lifts and corresponding cam angles of the input data the velocity $\dot{l}_f$ and the acceleration $\ddot{l}_f$ of the follower are estimated using the numerical differentiation. Then V_c, V_{fm}, and R_c are calculated from equations (2.1), (2.2), and (2.3) respectively.

9.4 The analysis of the hydraulic lash adjuster

The details of the analysis of the hydraulic lash adjuster and the solution procedure have been given in Chapter 3. With known displacement z of the plunger relative to the follower, the relative velocity $\dot{z}$ and the acceleration $\ddot{z}$ of the plunger are calculated by numerical differentiation.

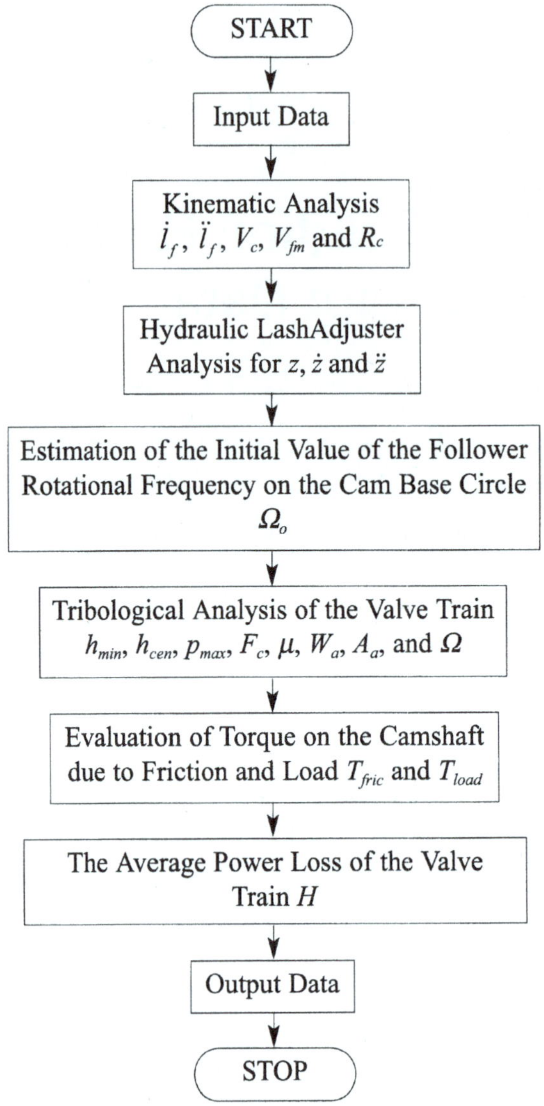

Fig. 9.1 The flow chart of the overall solution procedure for the tribological analysis of the valve train

9.5 Estimating the initial value of the follower rotational frequency

In order to predict the rotational frequency of the follower an initial rotational frequency Ω_o must be evaluated before starting the tribological analysis. As described in Chapter 8, the initial value is estimated from equation (8.1) by assuming the angular acceleration to be zero on the cam base circle a quarter of a revolution ahead of the cam nose. This means that

Table 9.1 Input data for the tribological analysis of the valve train

Cam and camshaft	
Cam lifts and corresponding cam angles, $l_c(\phi)$	mm
Cam width, C_w	m
Radius of the cam base circle, R_b	m
Cam taper, α'	degree
Poisson's ratio of cam material, υ_c	–
Young's modulus of cam material, E_c	Pa
Rotational frequency of the camshaft, ω	r/min
Follower and follower bush	
Diameter of the follower, D_f	m
Length of the follower, L_f	m
Moment of inertia of the follower, J	N m s^2/rad
Mass of the follower, m_f	kg
Radius of curvature of domed follower surface, R_f	m
Length of lower section of the follower, B_1	m
Length of upper section of the follower, B_2	m
Radial clearance between the follower and bush, C_f	m
Young's modulus of follower material, E_f	Pa
Poisson's ratio of follower material, υ_f	–
Hydraulic lash adjuster	
Radial clearance between the plunger and bush, h_1	m
Radial clearance between the plunger and cylinder, h_2	m
Overlap between the plunger and the bush, H_1	m
Overlap between the plunger and the cylinder, H_2	m
Stiffness of the plunger spring, k_{ps}	N/m
Mass of the plunger, m_p	kg
Inlet oil pressure to hydraulic lash adjuster, p_o	Pa
Radius of the plunger, r_1	m
Radius of the plunger cylinder, r_2	m
Radius of the curvature of domed plunger end, R_p	m
Precompression of the plunger spring, δ_p	m
Radius of check valve orifice, r_o	m
Flow loss coefficient of check valve orifice, ξ	–
Valve, valve spring, valve stem and valve guide	
Diameter of the valve stem, D_v	m
Length of the valve guide, L_v	m
Radial clearance between the valve stem and guide, C_v	m
Stiffness of the valve spring, K_{vs}	N/m
Mass of the valve, the spring collar and clips, m_v	kg
Mass of the valve spring, m_s	kg
Precompression of the valve spring, δ	m
Surface roughness	
Composite roughness height, $\sigma = \sqrt{\sigma_c^2 + \sigma_f^2}$	m
Product of asperity density, asperity radius of curvature and composite roughness height, $\eta\beta\sigma$	–
Ratio of the asperity radius of curvature to the composite roughness height, $\sqrt{\sigma/\beta}$	–
Asperity sliding coefficient of friction, μ_a	–
Lubricant properties	
Viscosity–pressure coefficient of the lubricant, α	m^2/N
Dynamic viscosity of the lubricant at atmospheric pressure, η_o	Pa s
Density of the lubricant, ρ	kg/m^3
Rate of change of shear stress with pressure, γ	–
The Eyring stress of the lubricant, τ_o	Pa
Limiting shear stress of the lubricant, τ_L	Pa

$$T_{fc}(\Omega_o)=T_{fb}+T_{fp}=\frac{\Omega_o D_f^2\left(F_{f1}+F_{f2}\right)}{4i_f}+\frac{3\pi\mu_c aP}{16} \tag{9.1}$$

It is known that T_{fc} is an implicit function of Ω_o. Thus the bisection method is adopted to solve this equation. An iterative loop is required in the solution procedure.

9.6 The tribological analysis of the valve train

The flow chart of the main sub-routine for the tribological analysis of the valve train is presented in Fig. 9.2. The sub-routine starts with initializing the follower rotational frequency to zero at each cam angle except for the first one which is given the initial value Ω_o. The solution is carried out step by step with one degree cam angle increment for the whole cam cycle. As the variation of the coefficient of friction is continuous, for each cam angle the initial coefficient of friction is taken to be the one at the previous cam angle.

The details for predicting the cam load, the maximum Hertzian stress, the rotational frequency of the follower, the minimum and the maximum film thicknesses, the friction, and the friction coefficient at each cam angle have been given in corresponding sections.

For each cam angle the current follower rotational frequency is compared with the one predicted in the previous cam cycle. If the relative difference is less than the tolerance the analysis is completed. As the array of Ω is initialized to zero before the solution loop of the cam cycle, convergent results can only be expected after the completion of the analysis on a full cam cycle.

9.7 The friction torque of the camshaft and average power loss

Knowing the friction arising from the cam/follower interface the friction torque on the camshaft is predicted by equations (7.8), (7.9), and (7.10). The integration of equation (7.11) gives the average power loss of the valve train. In the current analysis one cam cycle is equally divided into 360 steps with one degree cam angle increment. So the average power loss can be approximated by the following equation:

$$H=\frac{\omega}{n}\left[\sum_{i=2}^{n}\left(T_{fric}\right)i+\frac{\left(T_{fric}\right)_1}{2}+\frac{\left(T_{fric}\right)_n}{2}\right] \tag{9.2}$$

9.8 The output data

In order to present the variation of some important parameters along the cam cycle the results of the analysis are output as a function of cam angle. For each cam angle the output parameters are listed in Table 9.2. In spite of this the average power loss of the valve train and the average rotational frequency of the follower in one cam cycle are also provided.

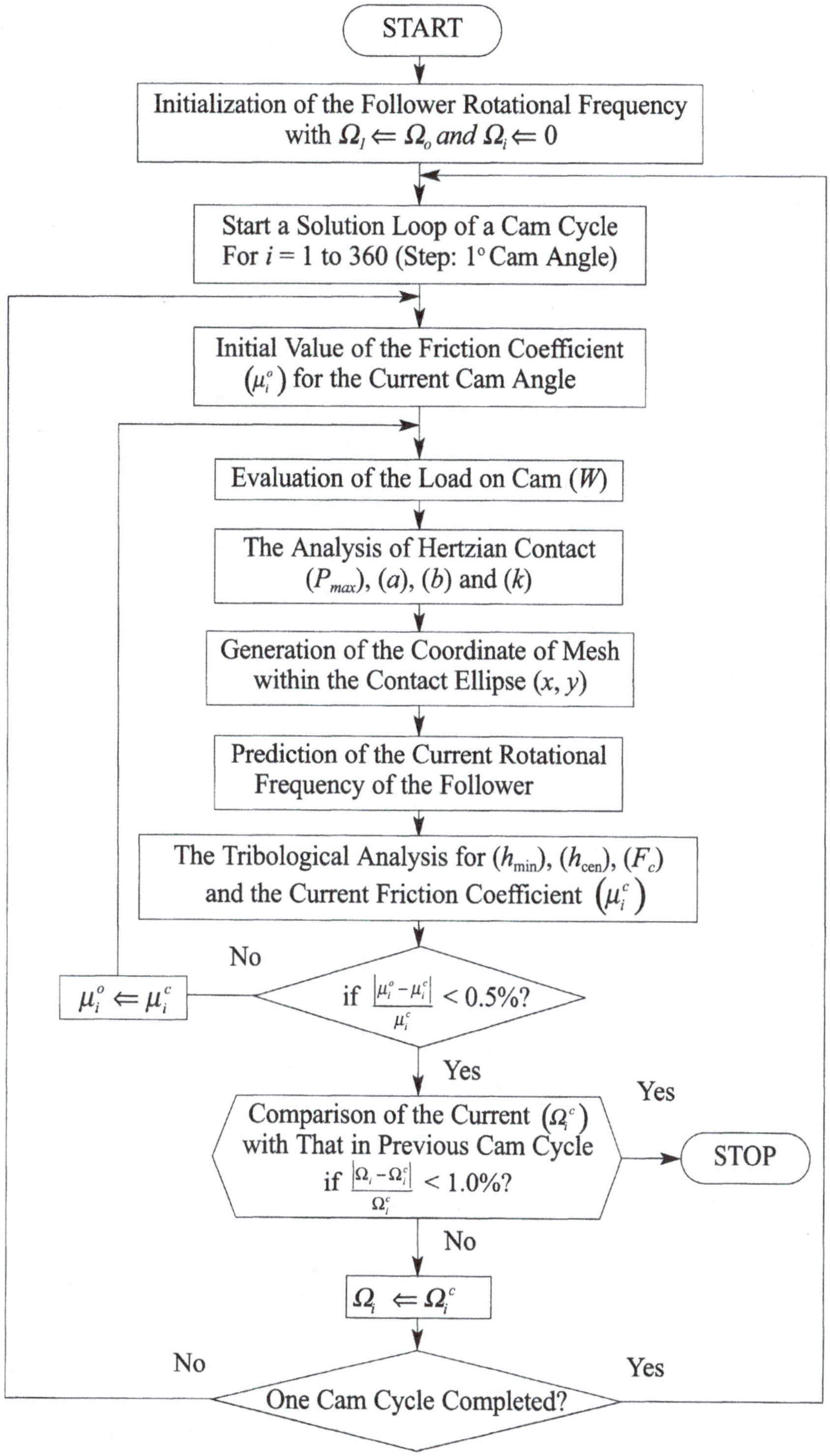

Fig. 9.2 The flow chart of the main sub-routine for the tribological analysis of the valve train

It should be pointed out that the current output follower velocity, the acceleration, and the velocities of contact point relative to the cam and relative to the follower are normalized by the camshaft rotational frequency. So these variables are dependent only upon the cam profile. It can be seen that the first nine variables listed in the table are only related to the kinematic analysis of the valve train. Thus, for a given cam and follower pair, the program generates only one output data file containing these variables for different operating conditions.

Table 9.2 Output data from the tribological analysis of the valve train

Cam angle, ϕ	degree
Cam lift, l_c	m
Normalized velocity of the follower, $\dot{l}_f / \omega$	m
Normalized acceleration of the follower, $\ddot{l}_f / \omega^2$	m
Normalized velocity of contact point relative to the cam, V_c/ω	m
Normalized velocity of contact point relative to the stationary follower, V_{fm}/ω	m
Radius of curvature of the cam, R_c	m
Effective radius of curvature, R	m
Eccentricity of the contact point from the camshaft centre, e	m
Load on cam, W	N
Maximum Hertzian pressure, p_{max}	Pa
Semi-major axis of ellipse of the contact, a	m
Ellipticity ratio of the semi-major axis to semi-minor axis, k	–
Effective entraining velocity of the lubricant, u_e	m/s
Angle between lubricant entrainment vector and minor axis, θ	degree
Asperity load, W_a	N
Friction due to asperity contact, F_a	N
Minimum lubricant film thickness, h_{min}	m
Central lubricant film thickness, h_{cen}	m
Friction of the cam/follower interface, F_c	N
Coefficient of friction of the cam/follower interface, μ_c	–
Friction torque of the valve train, T_{fric}	N m
Torque due to the cam load, T_{load}	N m
Plunger displacement relative to the follower, z	m
Rotational frequency of the follower, Ω	rad/s
Average power loss of the valve train, H	N m/s
Average rotational frequency of the follower, $\bar{\Omega}$	rad/s

Chapter 10

An Example of the Tribological Analysis of a Valve Train

10.1 Introduction

An example of the tribological analysis of the cam and follower of the intake valve train of the ZETEC engine is presented in this chapter to demonstrate the function of the model and some details of the organization of the input data. The output parameters of the analysis are presented in the form of graphs which make it easy for designers to assess the tribological performance of the mechanism. A brief discussion of the results is also given in this chapter.

10.2 Parameters of the intake valve train of the ZETEC engine

The sample given in this section corresponds to the intake valve train of the 1.8/2.0 litre high-output performance ZETEC engine. All input data required for the analysis are listed in Table 10.1. The nominal dimensions of each component of the valve train adopted in the analysis are taken from the drawings of the engine provided by the Ford Motor Company. The radial clearance between two components is taken to be the mean value according to the tolerance limits.

The lubricant for the ZETEC engine is specified by the Ford Motor Company as ESSO 5W/30. The viscosity–temperature characteristics have been determined with a U-tube viscometer (type BS/IP/SL) at the Institute of Tribology, University of Leeds. According to the operating condition in the fired engine, the bulk temperature of the valve train is about 95.0 °C. Knowing the bulk temperature of the valve train, the lubricant viscosity η_o at atmospheric pressure is calculated from the empirical formula which is a best-curve fit to the calibrated data of the lubricant. As three rheological parameters (γ, τ_o, and τ_L) of the lubricant 5W/30 were not available at the time of the experiment, the parameters of mineral oil HVI 650 given by Evans and Johnson [1] are adopted for the present example.

It has been shown in reference [2] that for most rough surfaces the product of the asperity density, the asperity radius of curvature, and the composite roughness height $\eta\beta\sigma$ is a constant between 0.03 and 0.05 and the ratio of the asperity radius of curvature to the composite roughness height $\sqrt{(\sigma / \beta)}$ is also a constant between 0.0001 and 0.01. The experimental study [3] has indicated that many cam and follower pairs, whatever the initial composite surface roughness height, have similar roughness heights of about 0.2 μm after running-in.

Table 10.1 The input data for the tribological analysis of the intake valve train of the ZETEC engine (1.8/2.0 litre, high output)

Name	*Magnitude*	*Unit*
Cam		
Cam lists and corresponding cam angles, $l_c(\phi)$		mm
Cam width, C_w	11.0×10^{-3}	m
Radius of the cam base circle, R_b	18.0×10^{-3}	m
Cam taper, α'	0.0167	degree
Young's modulus of cam material, E_c	170×10^9	Pa
Poisson's ratio of cam material, υ_c	0.28	–
Follower and follower bush		
Diameter of the follower, D_f	28.4×10^{-3}	m
Length of the follower, L_f	26.5×10^{-3}	m
Moment of inertia of the follower, J	6.8423×10^{-6}	N m s^2/rad
Mass of the follower, m_f	54.0×10^{-3}	kg
Radius of curvature of domed follower surface, R_f	8.0	m
Length of lower section of the follower, B_1	9.65×10^{-3}	m
Length of upper section of the follower, B_2	14.5×10^{-3}	m
Radial clearance between the follower and bush, C_f	24.0×10^{-6}	m
Young's modulus of follower material, E_f	204×10^9	Pa
Poisson's ratio of follower material, υ_f	0.29	–
Hydraulic lash adjuster		
Radial clearance between the plunger and bush, h_1	15.25×10^{-6}	m
Radial clearance between the plunger and cylinder, h_2	7.5×10^{-6}	m
Overlap between the plunger and the bush, H_1	6.50×10^{-3}	m
Overlap between the plunger and the cylinder, H_2	7.25×10^{-3}	m
Stiffness of the plunger spring, k_{ps}	1.0×10^4	N/m
Mass of the plunger, m_p	8.5×10^{-3}	kg
Inlet oil pressure to hydraulic lash adjuster, p_o	3.0×10^5	Pa
Radius of the plunger, r_1	6.990×10^{-3}	m
Radius of the plunger cylinder, r_2	5.625×10^{-3}	m
Radius of the curvature of domed plunger end, R_p	0.1456	m
Precompression of the plunger spring, δ_p	1.8×10^{-3}	m
Radius of check valve orifice, r_o	1.41×10^{-3}	m
Flow loss coefficient of check valve orifice, ξ	0.625	–
Valve, valve spring, valve stem, and valve guide		
Diameter of the valve stem, D_v	6.043×10^{-3}	m
Length of the valve guide, L_v	36.0×10^{-3}	m
Radial clearance between the valve stem and guide, C_v	20.0×10^{-6}	m
Stiffness of the valve spring, K_{vs}	37.634×10^3	N/m
Mass of the valve, the spring collar, and clips, m_v	60.0×10^{-3}	kg
Mass of the valve spring, m_s	35.0×10^{-3}	kg
Precompression of the valve spring, δ	5.5801×10^{-3}	m
Surface roughness		
Composite roughness height, $\sigma = \sqrt{\sigma_c^2 + \sigma_f^2}$	0.2×10^{-6}	m
Product of asperity density, asperity radius of curvature, and composite roughness height, $\eta\beta\sigma$	0.05	–
Ratio of the asperity radius of curvature to the composite roughness height, $\sqrt{(\sigma / \beta)}$	1.0×10^{-3}	–
Asperity sliding coefficient of friction, μ_a	0.2	–
Lubricant properties		
Viscosity–pressure coefficient of the lubricant, α	0.20×10^{-7}	m^2/N
Dynamic viscosity of the lubricant at atmospheric pressure, η_o	9.865×10^{-3}	Pa. s
Density of the lubricant, ρ	900.0	kg/m^3
Rate of change of shear stress with pressure, γ	0.08	–
Eyring stress of the lubricant, τ_o	8.0×10^6	Pa
Limiting shear stress of the lubricant, τ_L		Pa
Operating conditions of the valve train		
Temperature of the operation of the valve train	95.0	°C
Rotational frequency of the camshaft, ω	400–3000	r/min

The spring rate is obtained from a linear fit to two data pairs of spring compressions and spring loads specified in the valve spring drawing. It should be pointed out that the precompression of the valve spring listed in Table 10.1 is not the initial compression of the valve spring when it is assembled into the cylinder head. It is, in fact, an equivalent value relating to the ratio of the initial load on the valve spring at zero cam lift to the spring stiffness.

In order to investigate the effects of the camshaft rotational frequency on the tribological performance of the valve train, the analysis was performed with the camshaft rotational frequency at 400, 750, 1000, 1500, 2000, and 3000 r/min respectively. This corresponds to the variation of the engine rotational frequency from 800 to 6000 r/min. For convenience sake the unit of revolutions per minute (r/min) is quoted for the rotational frequency in the remainder of this chapter.

10.3 Results of the tribological analysis

In order that a quick assessment of the tribological performance of the valve train can be made the results are plotted as a function of the cam angle in Figs 10.1 to 10.20 (see end of chapter for all figures). In addition, the average follower rotational frequency, the average friction torque, and the average power loss are presented as a function of the camshaft rotational frequency in Figs 10.21, 10.22, and 10.23 respectively. The graphics shown in this chapter were produced with the UNIRAS V6.1f package on a SUN workstation.

10.4 A brief discussion of the results

The kinematic parameters which are only dependent upon the geometry of the cam and the follower are shown in Figs 10.1 to 10.5. As indicated in the previous chapters the velocity and the acceleration of the follower were normalized by the camshaft rotational frequency. Thus, the unit for these two parameters should be m/rad and m/rad^2 respectively (see Figs 10.2 and 10.3). The eccentricity of the contact point from the centre of the camshaft is presented in Fig. 10.5.

The parameters relating to the load and the Hertzian contact at different camshaft rotational frequencies are presented in Figs 10.6 to 10.8.

The cases at different camshaft rotational frequencies of 400, 750, 1000, 1500, 2000, or 3000 r/min are identified by different line styles. The variation of the length of the major axis of the contact ellipse at different camshaft rotational frequencies is presented in Fig. 10.8. If the length is greater than the cam width the truncated contact condition will occur. This will cause a stress concentration.

The resultant entraining velocity of the lubricant is plotted as a function of the cam angle in Fig. 10.9. The angle between the lubricant entrainment vector and the minor axis of the contact ellipse is shown in Fig. 10.10. The influence of the rotation of the follower on the lubricant film thickness and friction at different portions of the cam cycle and at different camshaft rotational frequencies can be interpreted from this figure. This will help to explain some tribological characteristics due to the rotation of the follower.

Important parameters relating to the asperity contact, the lubricant film thickness, and the friction for evaluating the tribological performances of the cam and follower are plotted in Figs 10.11 to 10.16. In order to evaluate the contribution of the surface roughness, the friction arising from the asperity contact is presented as a percentage of the total friction of the cam/follower interface.

Figure 10.17 shows the cyclic variation of the friction torque of the valve train on the camshaft. The total torque arising from one valve train set is presented in Fig. 10.18. This gives a quick estimation of the maximum torque on the camshaft.

The plunger displacement relative to the follower in a cam cycle is illustrated in Fig. 10.19. The cyclic variations of the instantaneous rotational frequency of the follower are presented in Fig. 10.20. It can be seen that the peak of the rotational frequency is after the cam nose and shifts further away from the cam nose as the camshaft rotational frequency increases. An increase of the camshaft rotational frequency results in a reduction of the peak value of the follower rotational frequency.

The average follower rotational frequency is plotted in Fig. 10.21 against the camshaft rotational frequency. The increase of the camshaft rotational frequency results in an increase in the average follower rotational frequency initially, while a further increase of the camshaft rotational frequency from 2000 to 3000 r/min leads to a slight decrease of the average follower rotational frequency. This is due to the improvement of the lubrication condition within the cam/follower interface.

Figure 10.22 shows the variation of the average friction torque with the camshaft rotational frequency. The average friction torque decreases as the increase of the camshaft rotational frequency increases. The average power loss, however, increases as the camshaft rotational frequency increases (see Fig. 10.23).

10.5 References

1. **Evans, C. R.** and **Johnson, K. L.** (1986) The rheological properties of elastohydrodynamic lubricants. *Proc. Instn Mech. Engrs, Part C*, **200** (5), 625–633.
2. **Greenwood, J. A.** and **Tripp, J. H.** (1970–71) The contact of two nominally flat rough surfaces. *Proc. Instn Mech. Engrs*, **185**, 625–633.
3. **Harrison, P.** (1985) A study of the lubrication of automotive cams. PhD thesis, Mechanical Engineering Department, Leeds University, UK, October.

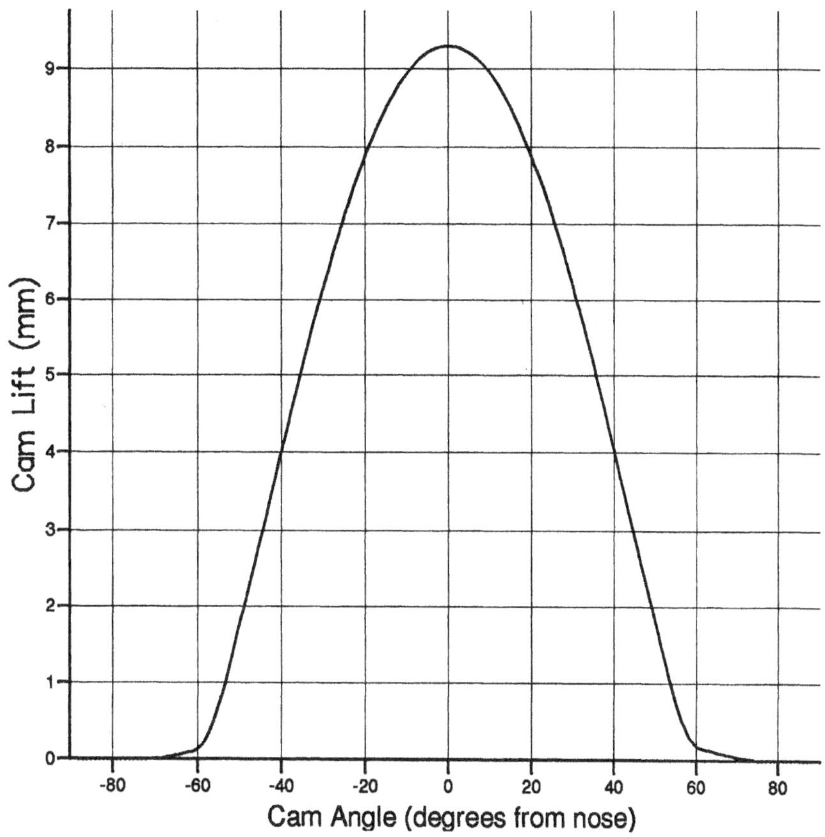

Fig. 10.1 The cam lift of the intake valve train (intake valve train of ZETEC engine (1.8/2.0 l) – design case)

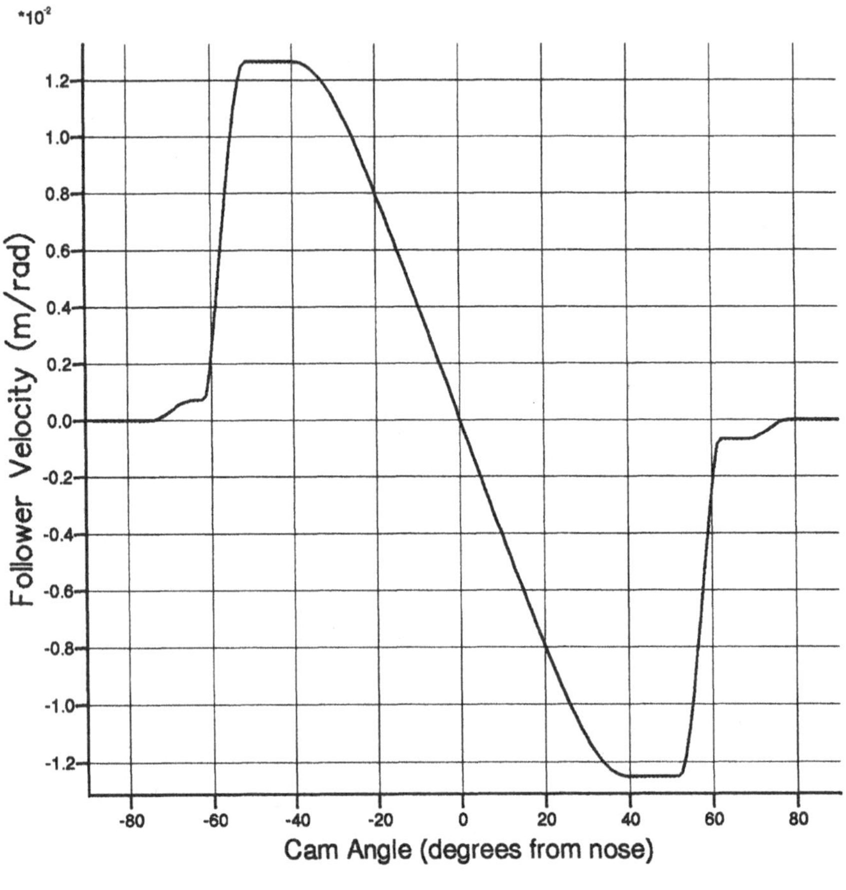

Fig. 10.2 The normalized velocity of the follower (intake valve train of ZETEC engine (1.8/2.0 l) – design case)

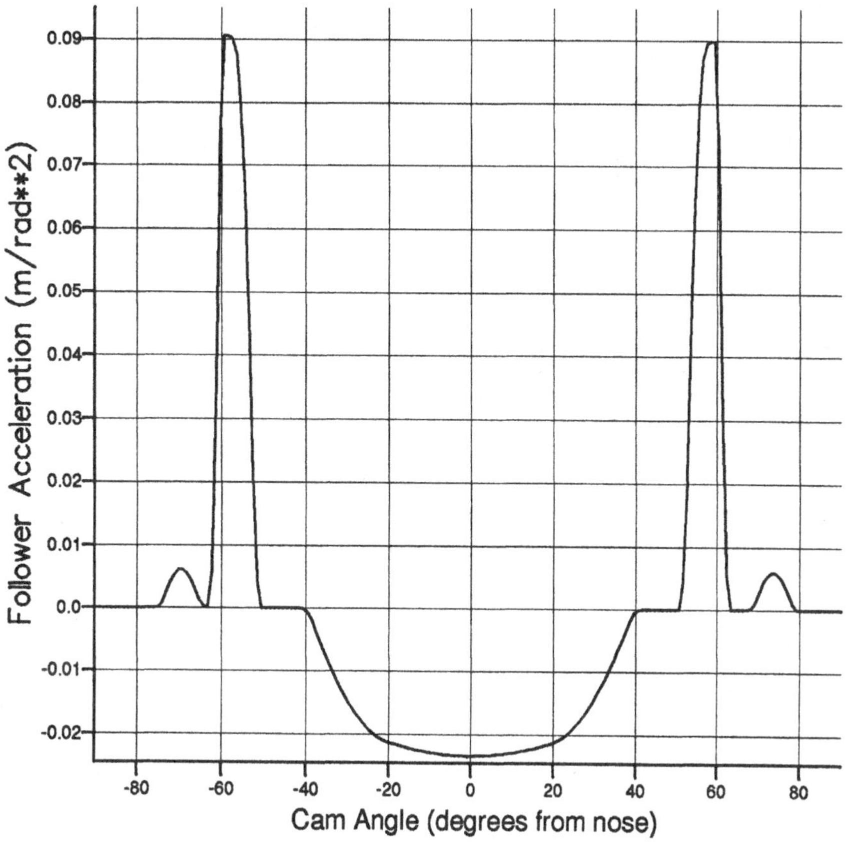

Fig. 10.3 The normalized acceleration of the follower (intake valve train of ZETEC engine (1.8/2.0 l) – design case)

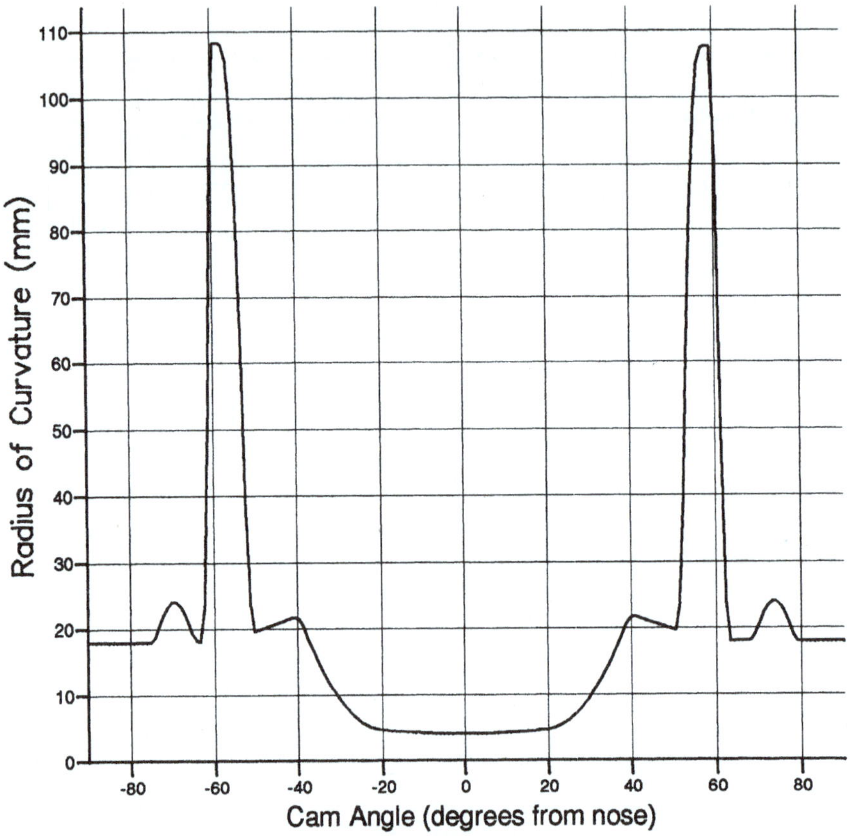

Fig. 10.4 The equivalent radius of curvature (intake valve train of ZETEC engine (1.8/2.0 l) – design case)

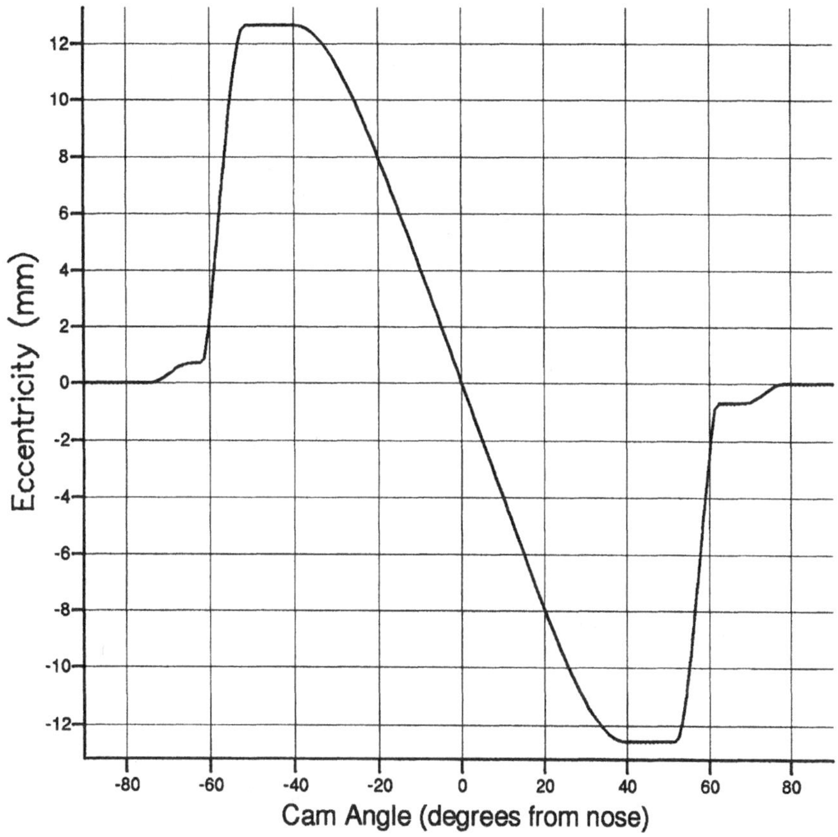

Fig. 10.5 The eccentricity of the contact point from the camshaft centre (intake valve train of ZETEC engine (1.8/2.0 l) – design case)

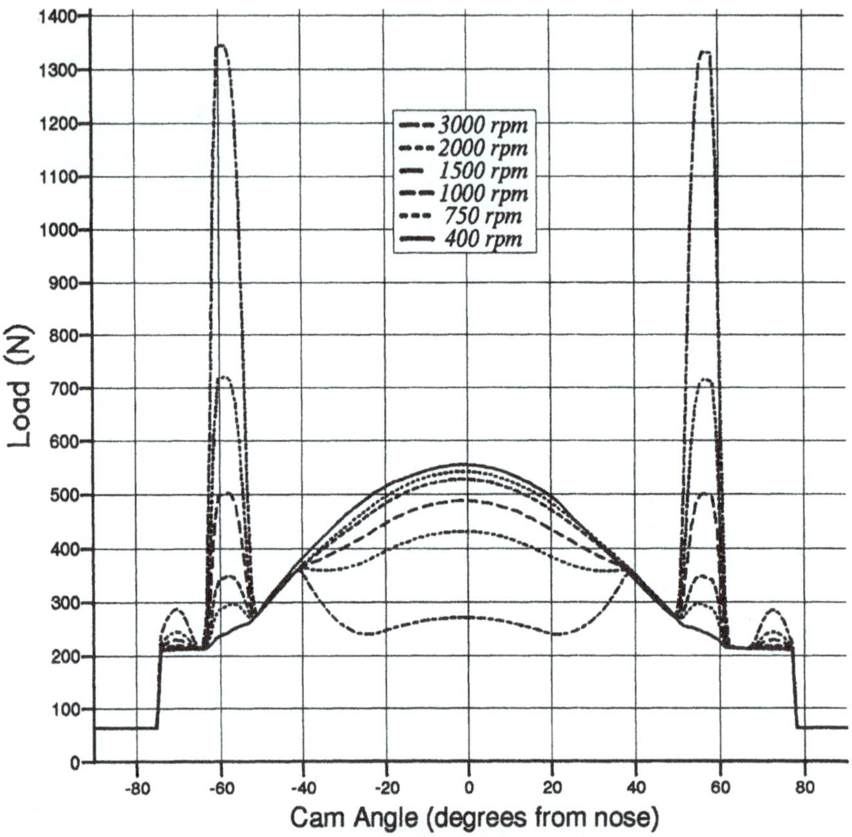

Fig. 10.6 The load on the cam (intake valve train of ZETEC engine (1.8/2.0 l) – design case)

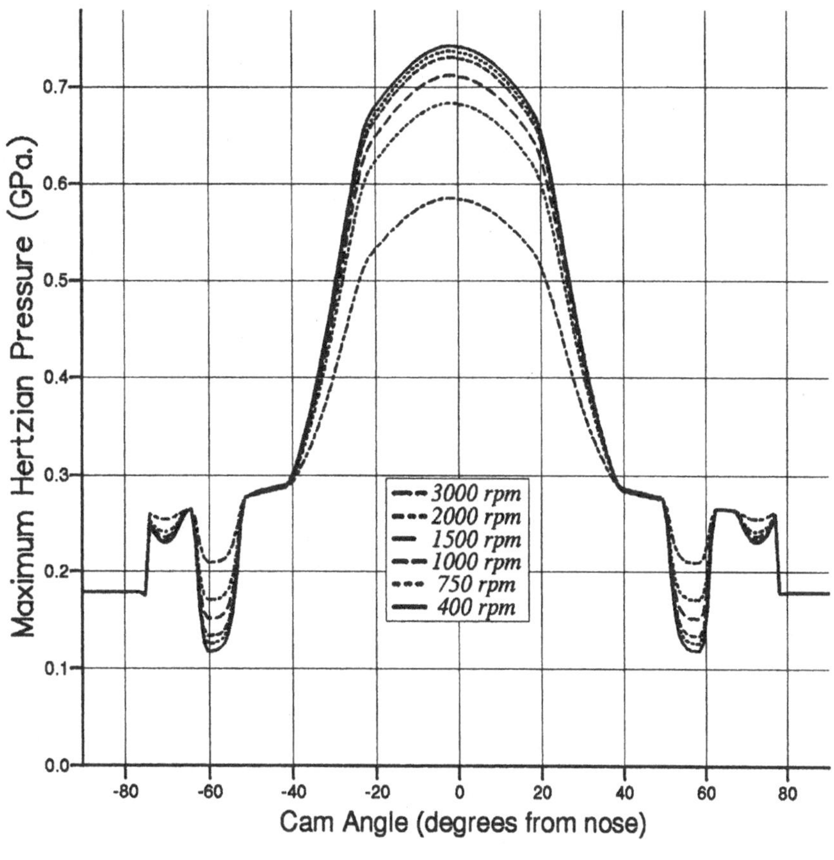

Fig. 10.7 The maximum Hertzian pressure (intake valve train of ZETEC engine (1.8/2.0 l) – design case, temperature 95°C)

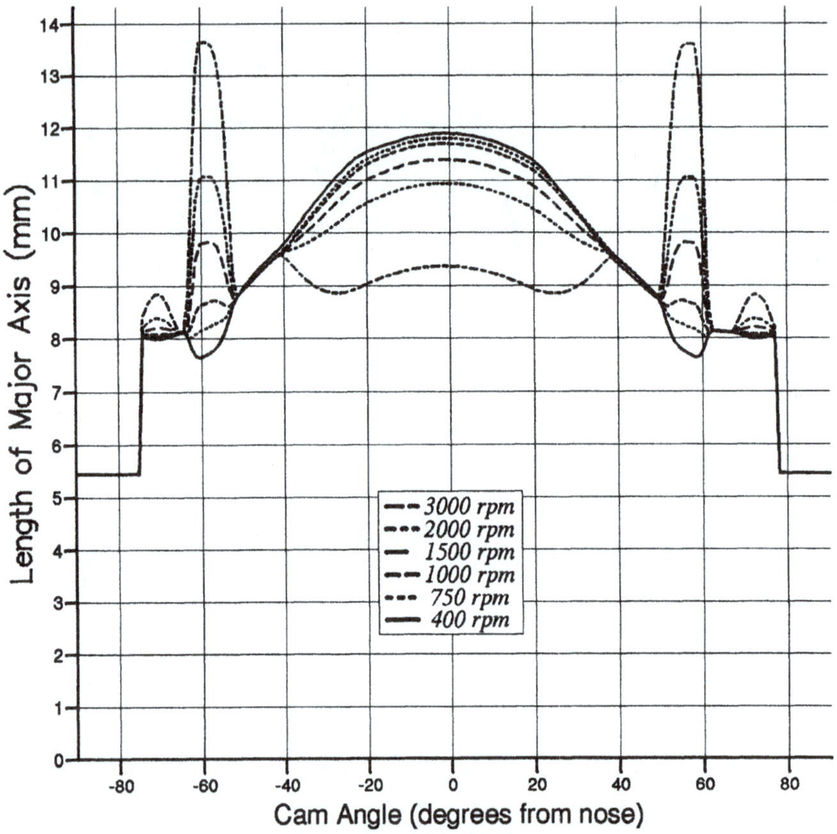

Fig. 10.8 The length of the major axis of the contact ellipse (intake valve train of ZETEC engine (1.8/2.0 l) – design case, temperature 95°C)

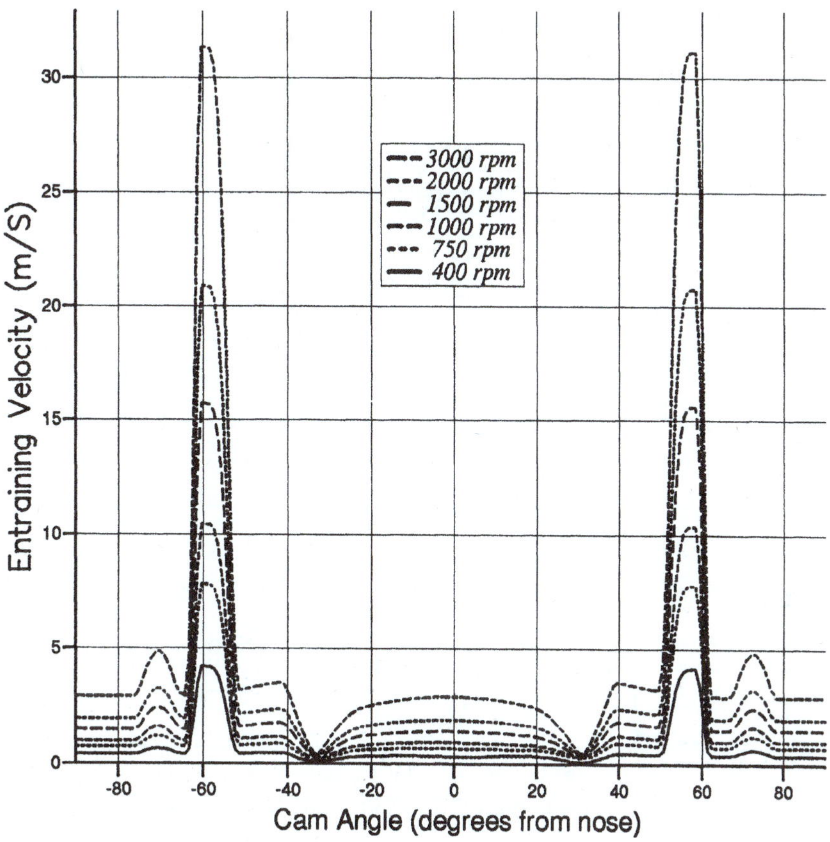

Fig. 10.9 The resultant entraining velocity of the lubricant (intake valve train of ZETEC engine (1.8/2.0 l) – design case, temperature 95°C)

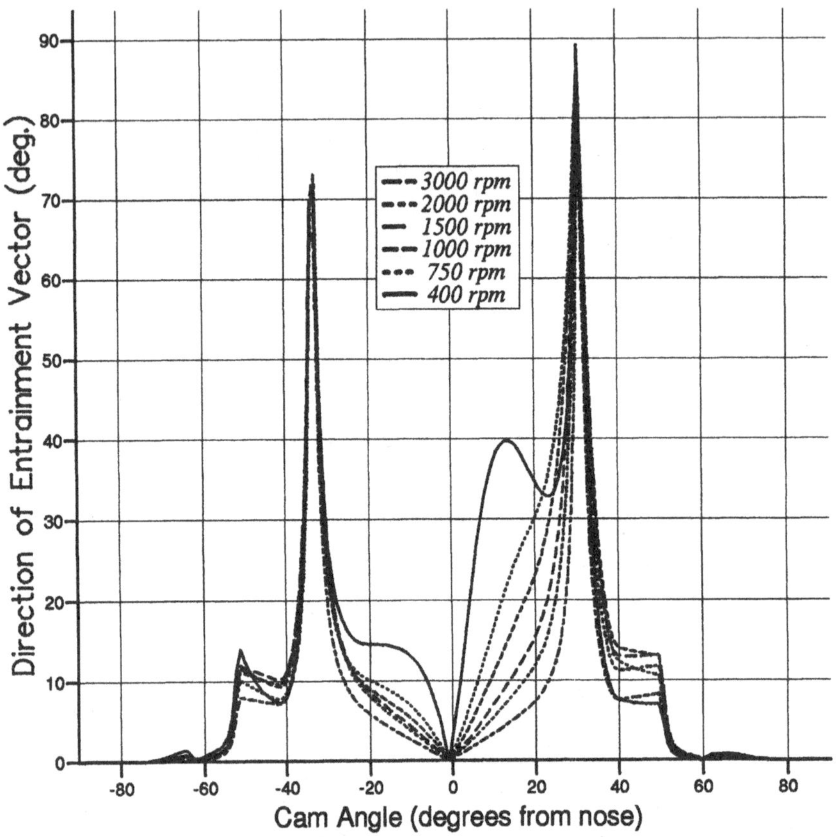

Fig. 10.10 The variation of direction of the lubricant entrainment vector (intake valve train of ZETEC engine (1.8/2.0 l) – design case, temperature 95°C)

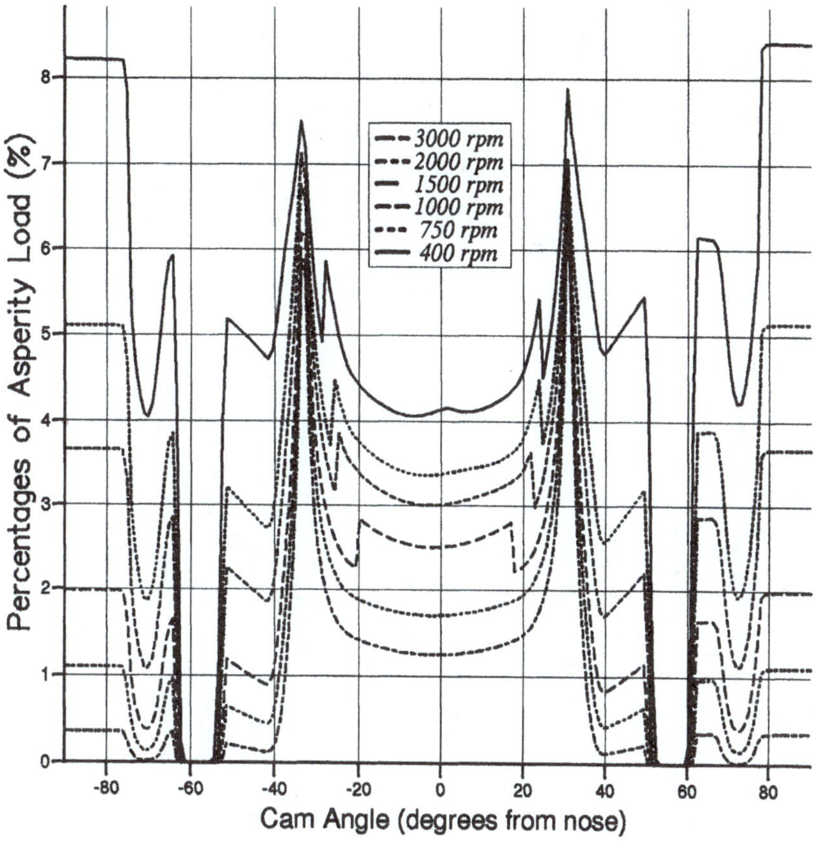

Fig. 10.11 The percentages of the load carried by asperities (intake valve train of ZETEC engine (1.8/2.0 l) – design case, temperature 95°C)

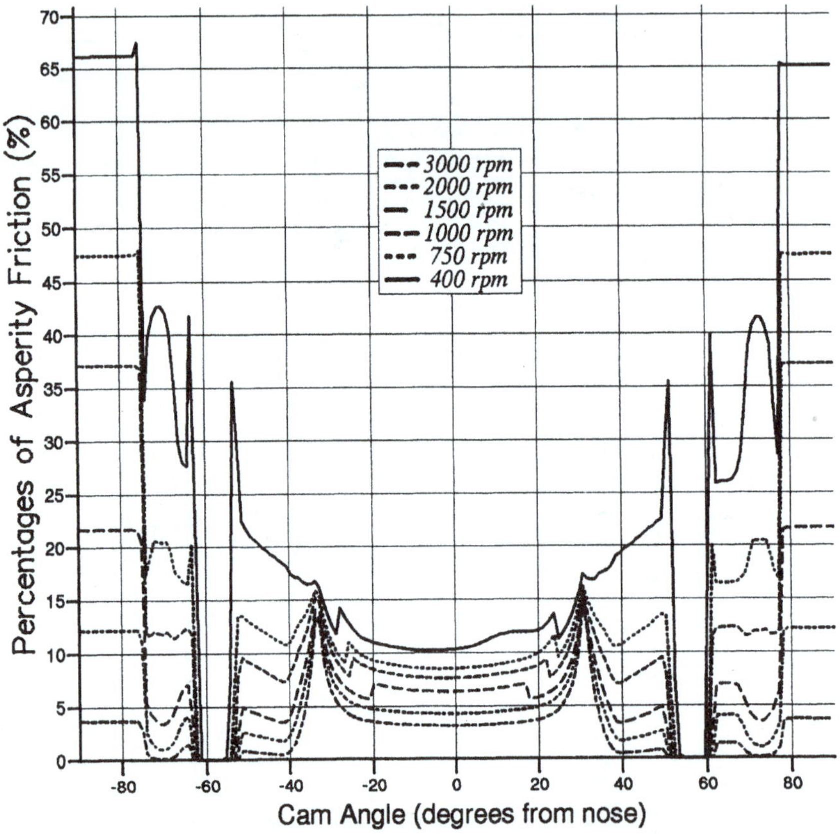

Fig. 10.12 The percentages of the friction due to asperity contact (intake valve train of ZETEC engine (1.8/2.0 l) – design case, temperature 95°C)

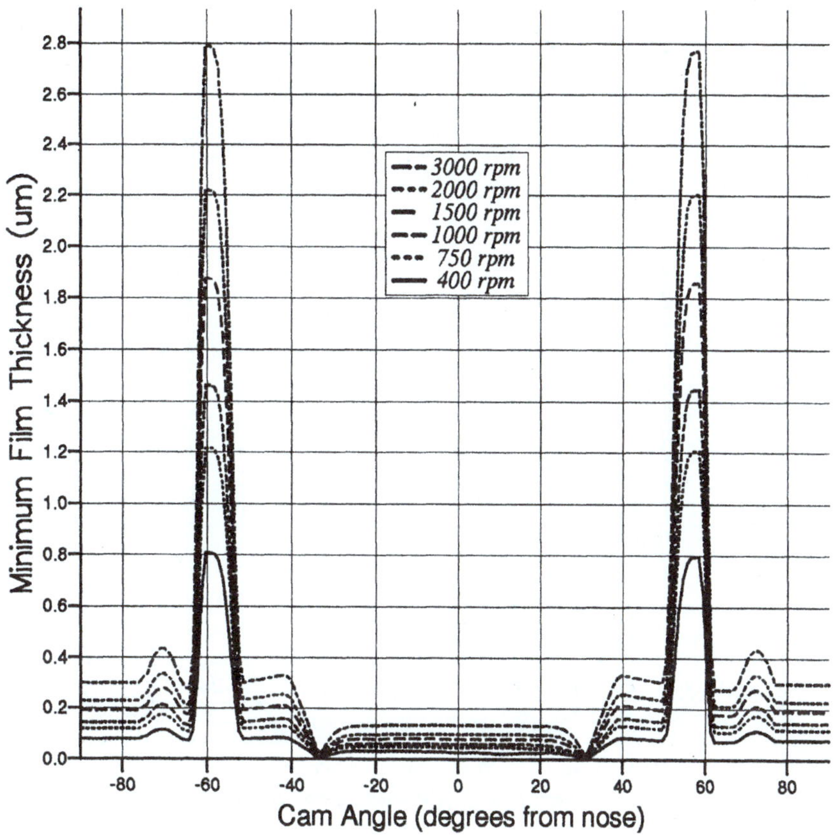

Fig. 10.13 The minimum lubricant film thickness (intake valve train of ZETEC engine (1.8/2.0 l) – design case, temperature 95°C)

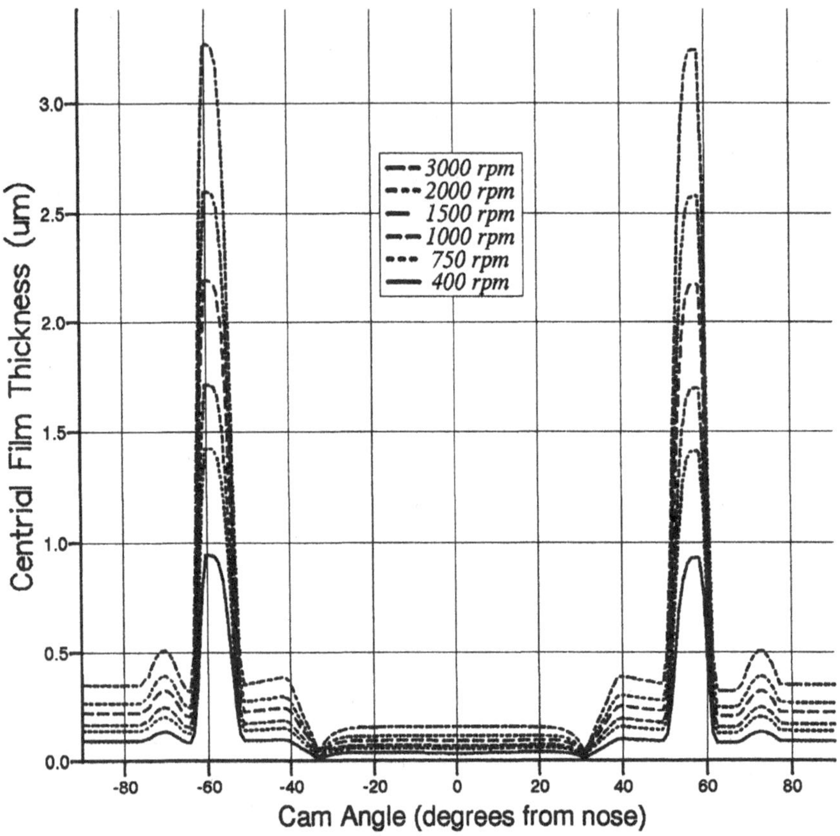

Fig. 10.14 The central lubricant film thickness (intake valve train of ZETEC engine (1.8/2.0 l) – design case, temperature 95°C)

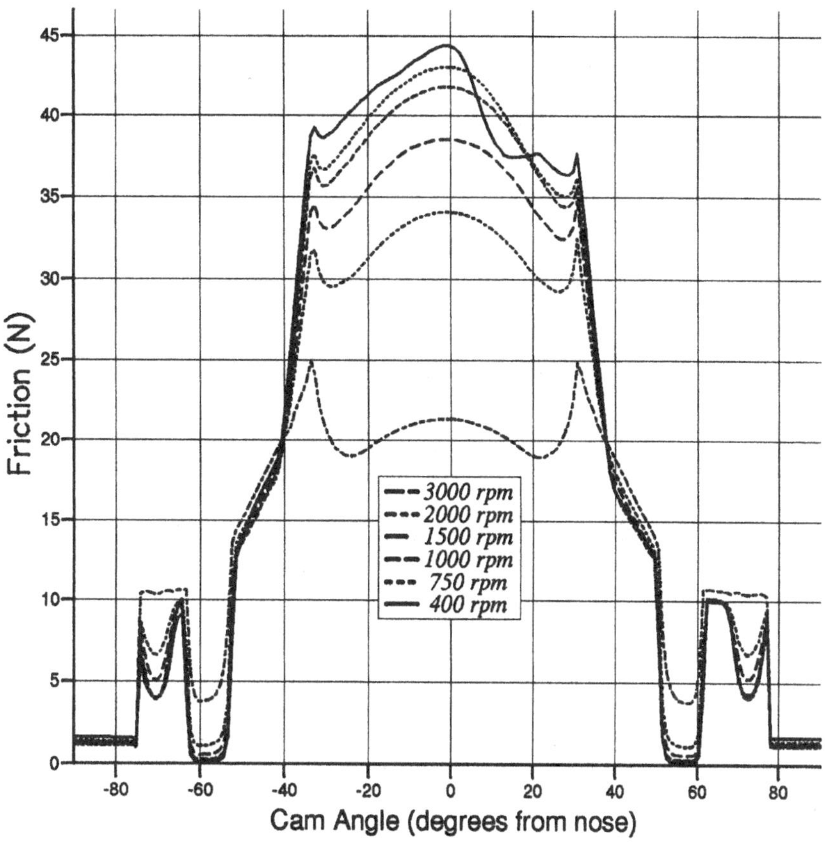

Fig. 10.15 The friction of the cam/follower interface (intake valve train of ZETEC engine (1.8/2.0 l) – design case, temperature 95°C)

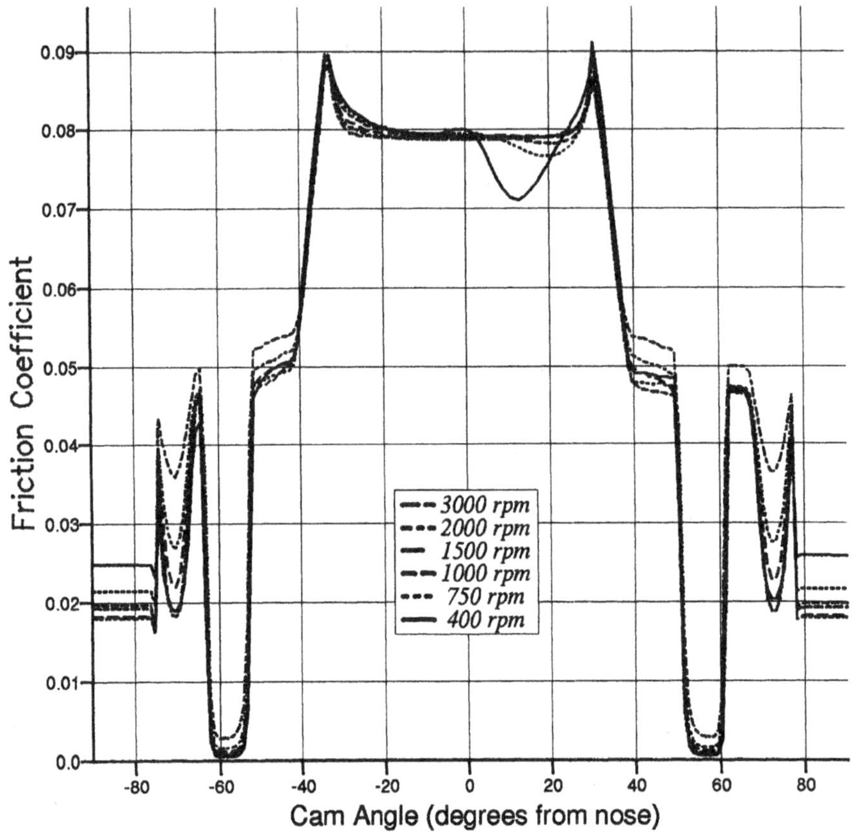

Fig. 10.16 The friction coefficient of the cam/follower interface (intake valve train of ZETEC engine (1.8/2.0 l) – design case, temperature 95°C)

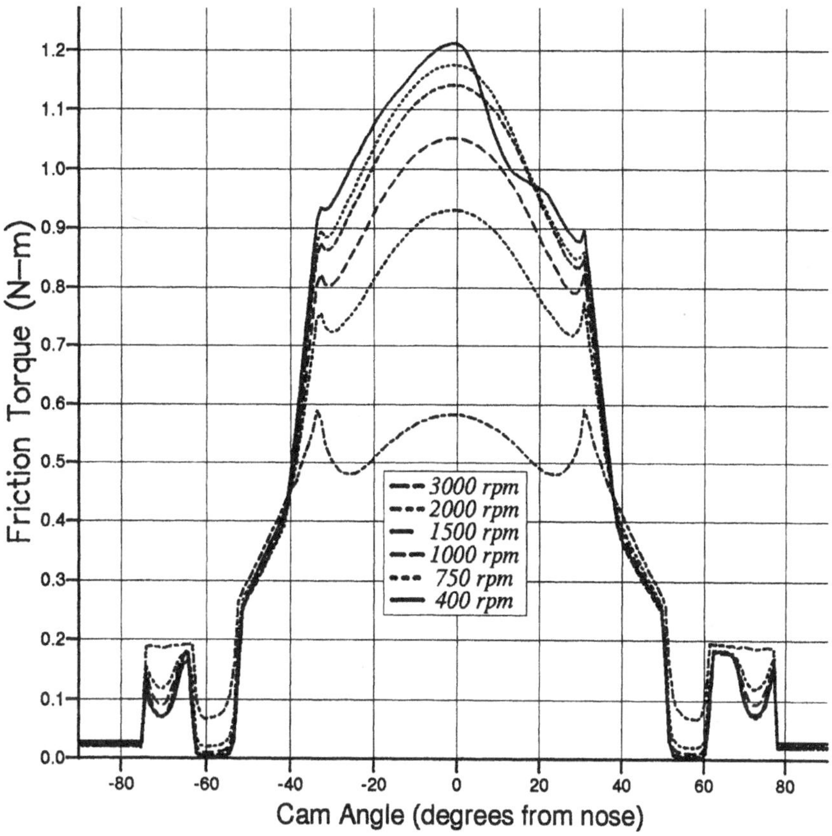

Fig. 10.17 The torque on the camshaft due to the cam/follower friction (intake valve train of ZETEC engine (1.8/2.0 l) – design case, temperature 95°C)

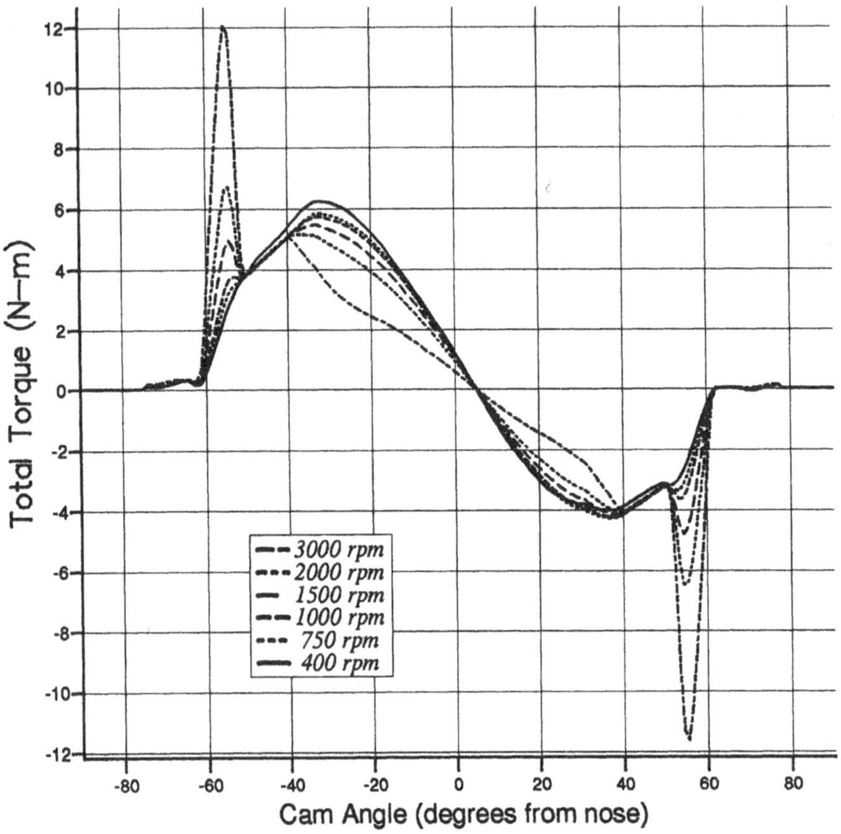

Fig. 10.18 The total torque on the camshaft (intake valve train of ZETEC engine (1.8/2.0 l) – design case, temperature 95°C)

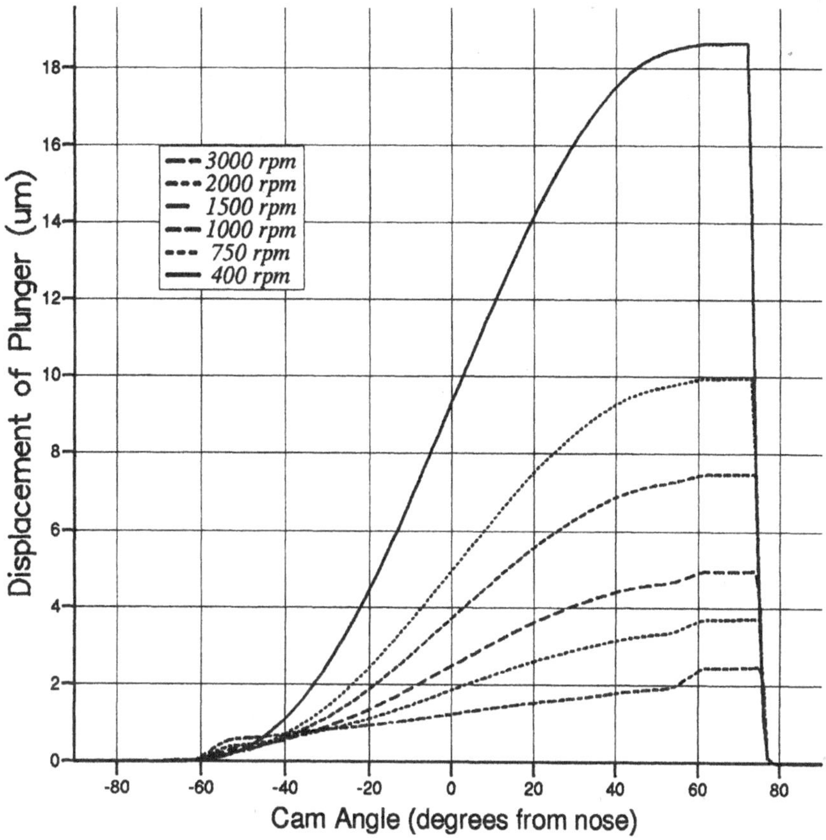

Fig. 10.19 The plunger displacement relative to the follower (intake valve train of ZETEC engine (1.8/2.0 l) – design case, temperature 95°C)

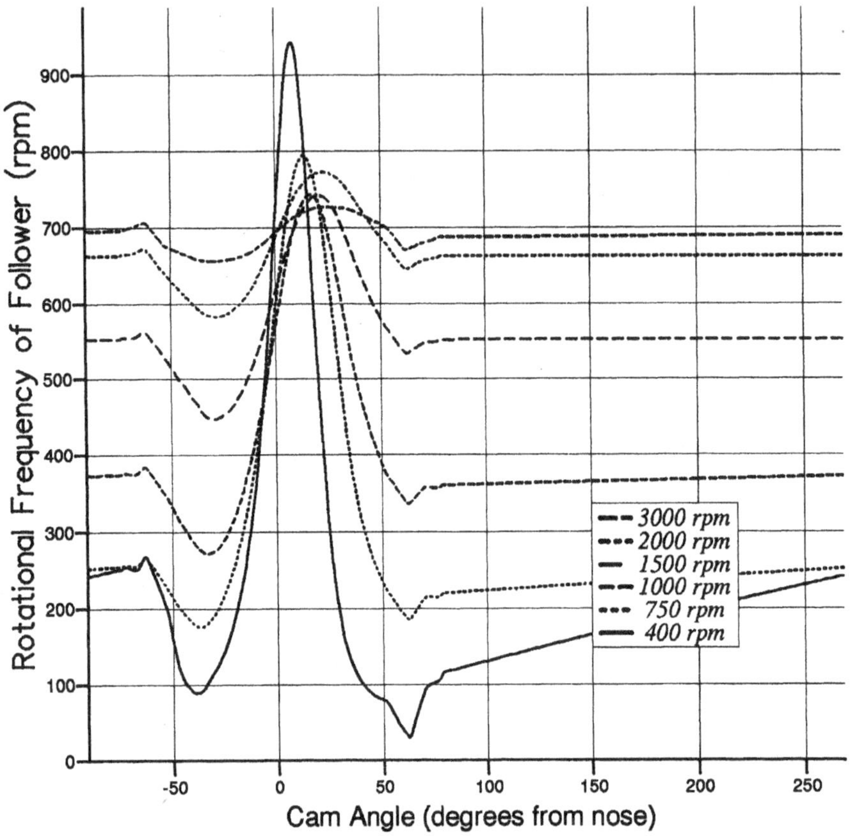

Fig. 10.20 The cyclic variation of the follower rotational frequency (intake valve train of ZETEC engine (1.8/2.0 l) – design case, temperature 95°C)

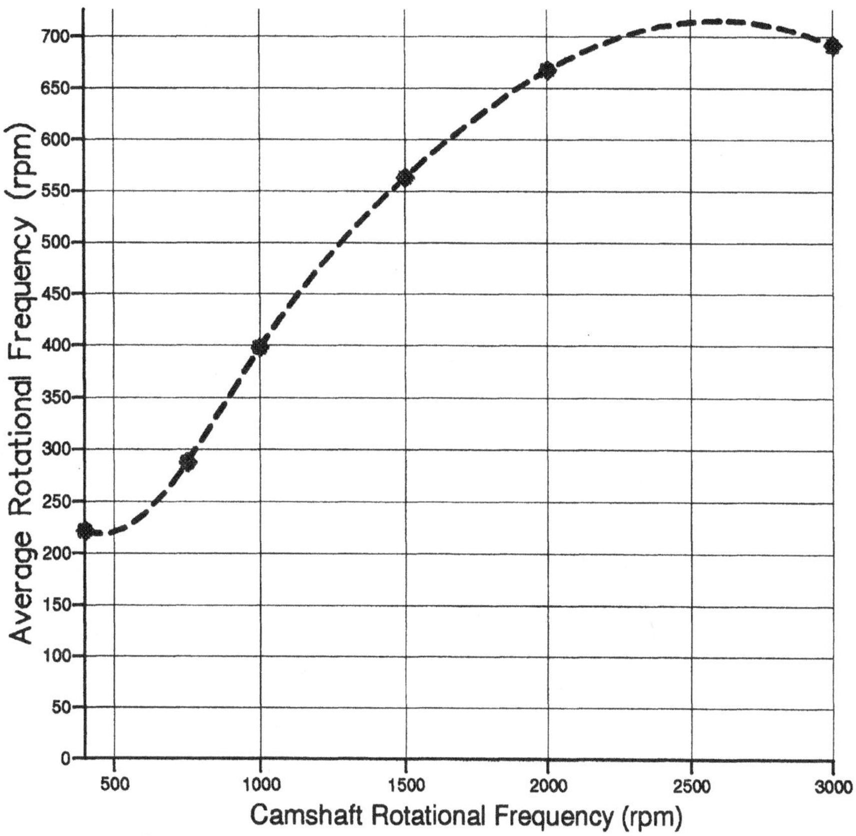

Fig. 10.21 The variation of the average follower rotational frequency (intake valve train of ZETEC engine (1.8/2.0 l) – design case, temperature 95°C)

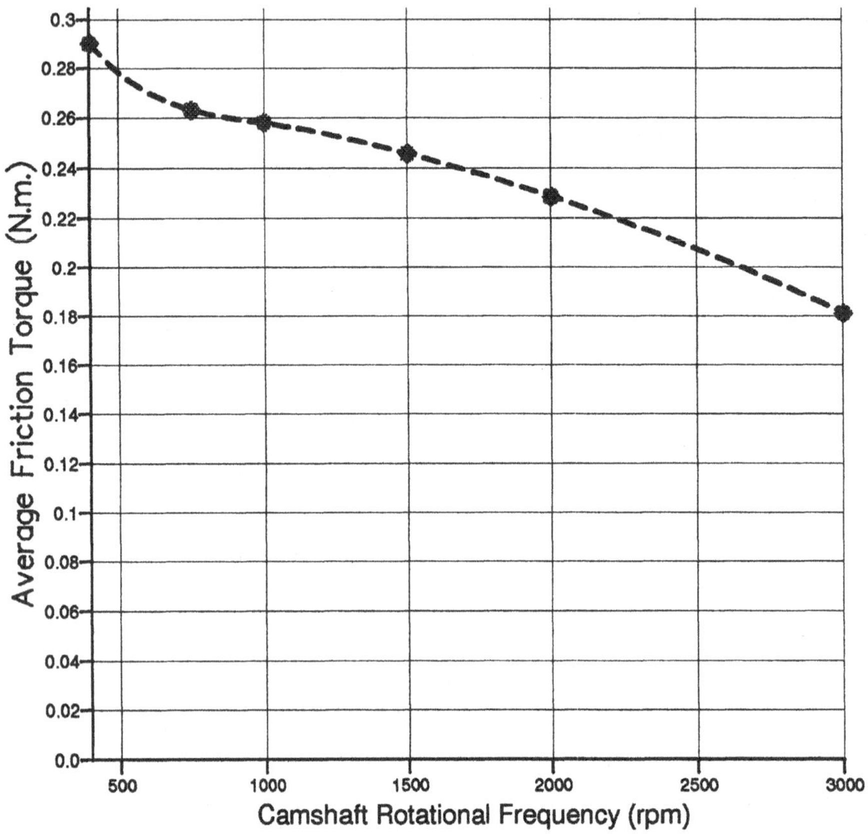

Fig. 10.22 The variation of the average friction torque of the valve train (intake valve train of ZETEC engine (1.8/2.0 l) – design case, temperature 95°C)

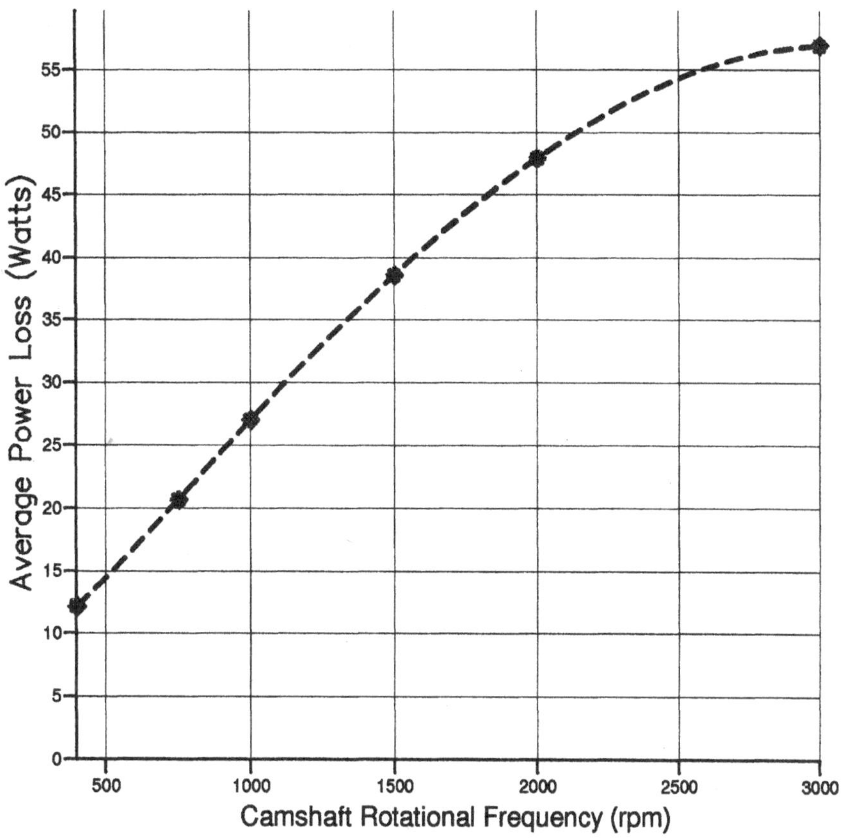

Fig. 10.23 The variation of the average power loss (intake valve train of ZETEC engine (1.8/2.0 l) – design case, temperature 95°C)

Part II – Experimental Study

Chapter 11

Introduction to Experimental Study

11.1 Preamble

In the second part of the book a detailed introduction to the experimental study of the valve train is given. This includes introduction to the test apparatus, the calibration and commissioning tests, data processing, test procedure, experimental measurements, and a brief discussion of the results.

A careful and comprehensive experimental programme consisting of a series of tests at different camshaft rotational frequencies and different bulk temperatures has been undertaken. This is to validate the theoretical model developed for predicting the tribological performance of the valve train having a taper cam and a domed follower with a hydraulic lash adjuster and to investigate the tribological performance of the valve train. Several important variables including the friction torque, the power loss, the follower rotational frequency, and the temperatures of the lubricant supply, inside the test cell and in the valve guide have been monitored in each test. Some sophisticated techniques for data acquisition and data processing, including a PC-based data acquisition system and a follower rotational frequency monitoring system, have been developed.

Two test procedures, the loop test and the duration test, have been adopted to examine the repeatability and consistency of the experimental measurement and to investigate the tribological performance of the valve train at different camshaft rotational frequencies. The results show that both test procedures produce the same results for the same operating conditions. This indicates that the test apparatus, the instrumentation, test procedures, and data processing developed in the current experimental study are very reliable and can be used for further studies on the tribological performance of a direct-acting valve train with different configurations.

Experimental data are presented in this book for torque, average friction torque, average power loss, and average rotational frequency of the follower as a function of camshaft rotational frequency, the cyclic variation of the friction torque, and the follower rotational frequency. These data clearly demonstrate the tribological performance of a valve train having an offset taper cam and a domed follower with hydraulic lash adjuster. The excellent overall correlation with theoretical predictions which is detailed in the next part of the book gives great confidence in dealing with novel and alternative designs.

11.2 The background of the current experimental study

As part of a long-term collaboration with the Ford Motor Company on studies of the automobile valve train, a comprehensive experimental programme examining the performance of a cam and a flat-faced bucket follower has already been undertaken. Full details have been recorded [1]. A comprehensive and carefully designed experimental programme consisting of a series of tests at different bulk temperatures and camshaft rotational frequencies was undertaken to investigate the tribological performance of the cam and follower. Several important features, including the friction torque, power loss, electrical resistivity across the contact of the cam and follower, and temperatures at the cam nose and in the valve guide were monitored in each test. By adopting an advanced technique for data acquisition and data processing which included an analogue to digital converter (ADC) data sampling system, a spectrum analyser, and a digital filter, and through undertaking a comprehensive dynamic analysis of the camshaft system, the instantaneous friction torque was successfully measured with the camshaft rotational frequency exceeding 2000 r/min. The agreement between the theoretical analysis and experimental measurement was encouragingly good. The details have been published in references [2] and [3].

The current experimental study of the valve train having a taper cam and a domed follower with hydraulic lash adjuster has been conducted on the same apparatus. An optical transmission technique is adopted to monitor the follower rotation. The top of the test cell has been redesigned to accommodate the hydraulic lash adjuster, an infra-red emitter, and an infra-red receiver (detector). A PC-based data acquisition system has been developed to provide a much higher sampling rate and longer sampling duration with much lower noise and distortion of the signal.

11.3 The aim of the experimental study

First, in order that the tribological performance of different configurations of a direct-acting valve train can be estimated, it is necessary to create a stable test apparatus and a reliable data acquisition system, to set up a standard test routine, and to develop a sophisticated technique for data processing. The current experimental study on the valve train is targeted to reach this objective.

Second, as has been described in the first part of the report, a sophisticated theoretical model has been developed to predict the tribological performance of an automobile valve train having a taper cam offset to a domed follower with hydraulic lash adjuster. It is believed that this is the first time that the follower rotation has been simulated by a theoretical model. In addition, the friction torque of the valve train might be affected due to follower rotation. Thus it is necessary to validate the model with experimental evidence which includes the measurement of the rotational frequency of the follower and the friction torque of the valve train.

Finally, the experimental study is to investigate the tribological performance of a specific direct-acting valve train which has an offset taper cam and a domed follower with a hydraulic lash adjuster. Further experimental study focused on the influence of some important variables, such as the camshaft rotational frequency and the bulk temperature, on the tribological performance of the valve train. This apparatus provides a unique facility to optimize the automobile valve train design.

11.4 References

1. **Zhu, G.** (1988) A theoretical and experimental study of the tribology of a cam and follower. PhD Thesis, Mechanical Engineering Department, Leeds University, UK, May.
2. **Dowson, D., Taylor, C. M.,** and **Zhu, G.** (1989) An experimental study of the tribology of a cam and flat-faced follower. 2nd International Conference on the *Combustion Engine – Reduction in Friction and Wear*, pp. 97–105.
3. **Dowson, D., Taylor, C. M.,** and **Zhu, G.** (1990) Experimental observations of lubricant film state between a cam and bucket follower using the electrical resistivity technique. *Proc. Jap. Int. Tribology Conference*, pp. 119–124.

Chapter 12

Test Apparatus and the Instrumentation

12.1 Introduction

The general view of the apparatus has been shown in Fig. 6.1 of reference [1]. In the present experimental study the main mechanical construction of the apparatus remains unchanged. Thus only major modifications to the apparatus and the instrumentation will be detailed in this section. This includes a redesigned top of the test cell, an independent oil supply system to the hydraulic lash adjuster, a data acquisition system, and the instrumentation for monitoring the follower rotation, sampling the camshaft torque, and locating the cam position.

12.2 The hydraulic lash adjuster

In order to accommodate a domed follower with a hydraulic lash adjuster, the top of the test cell has been redesigned. The details of the construction are presented in Fig. 12.1. As the hydraulic lash adjuster requires pressurized oil, a gear pump was adopted to pump the oil from the oil bath of the thermostatic circulator to the top of the test cell. A pressure gauge was inserted just before the oil inlet hole on the top to monitor the pressure of the oil being pumped into the hydraulic lash adjuster. The temperature of the inlet oil was measured by inserting a nickel/chromium thermocouple into the middle of the inlet oil pipe. This information will be used for estimating the lubricant viscosity between the follower and the bore.

12.3 The data acquisition system

In the previous experimental studies [1, 2] the analogue signal was sent to the ADC of the VAX/8600 computer which was installed about 100 metres away from the test apparatus. It was found that the long cable caused a severe distortion of the square wave signals due to the large capacitance. The maximum data that could be sampled in each operation was only about 32 000. In the current study the acquisition of high-quality digitalized square wave signals was essential to the monitoring of the rotational frequency of the follower.

To achieve this, a PC-based data acquisition system has been developed. Figure 12.2 is a line diagram of the instrumentation adopted in the current experimental study. This system includes an ELONEX/386s personal computer, a DMA 12-bit analogue-to-digital converter, and a comprehensive driver software coded in Microsoft C. The total

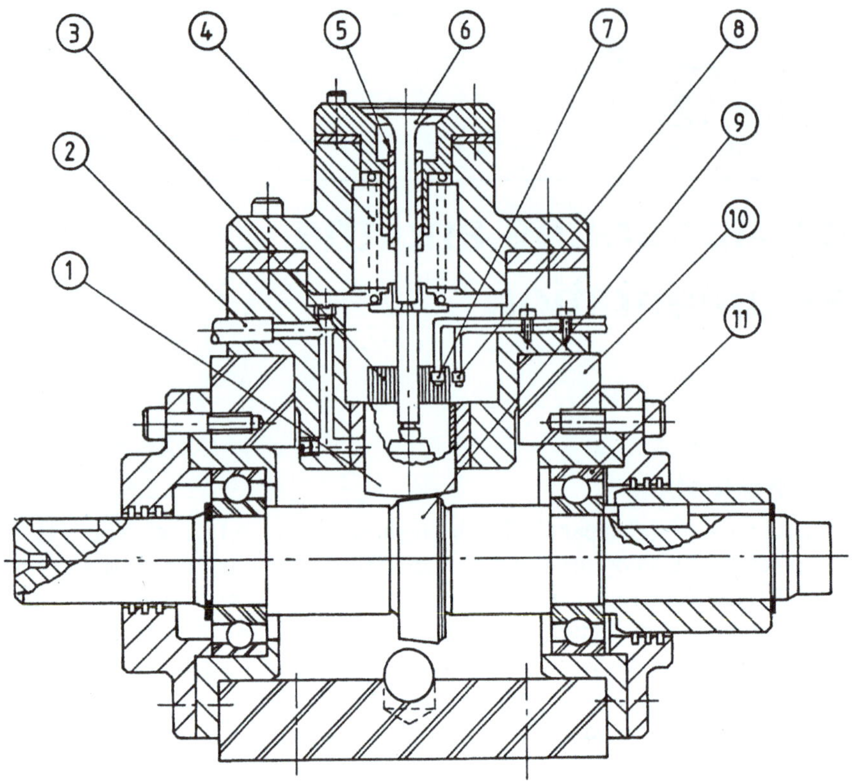

1	Follower	7	Infra-red emitter
2	Inlet oil to the hydraulic lash adjuster	8	Infra-red detector
3	Film	9	Cam
4	Valve spring	10	Test cell
5	Valve guide	11	Camshaft support bearing
6	Valve		

Fig. 12.1 Details of the test apparatus

number of sampling points was extended up to 300 000 with the maximum data acquisition rate being up to 200 kHz. The sampled signal of each channel can be displayed on the screen either in its original form or in post-processed form. If it is necessary the sampled data can be written into a data file on the hard disk of the computer for further processing.

A typical sampling run for 200 000 data points with 200 kHz sampling rate created a 1.22 Mbyte data file. This data file was then copied on to a 3.5 inch floppy disk and was transferred to an AMDAHL computer through the NOVELL SERVERS by the RAINBOW file transfer. The post-processing of the experimental data, including the data processing and producing hard copies of experimental measurements, was performed on the AMDAHL computer. The details of the data processing are given in Chapter 15.

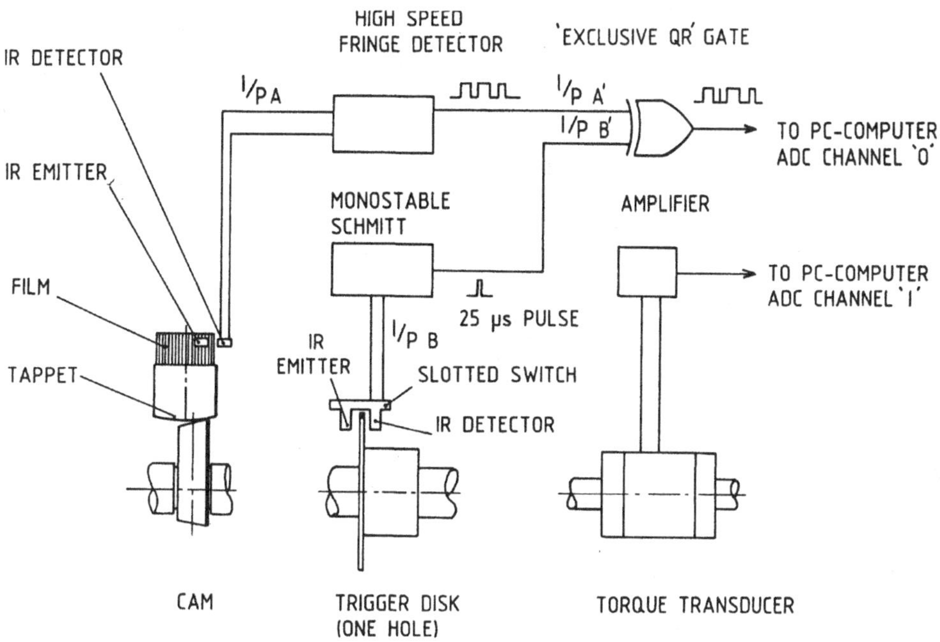

Fig. 12.2 A line diagram of the instrumentation for torque and follower rotational frequency measurement

12.4 Monitoring the follower rotation

Monitoring the follower rotational frequency is essential to the current experimental study and has proved to be the most difficult task. The most important point is that the basic construction of the valve train and the movement of the follower should not be affected by the monitoring system.

In order to measure the rotational frequency of the follower an optical transmission technique has been adopted. Figure 12.2 shows the detailed construction of the monitoring system which includes a 'sweet spot' type infra-red emitter and an infra-red receiver (or detector) with a switching frequency of up to 50 MHz. A thin circular film having 360 pairs of black and transparent stripes was inserted between the emitter and the detector. The film was glued on the top of the follower. Therefore, each cross of a pair of stripes to the emitter and the detector represented one degree angular displacement of the follower. The output current from the detector varied corresponding to each cross.

The infra-red detector was incorporated into a high-frequency photodiode fringe detection interface which consisted of a 'precision photodiode comparator' using an LM311 voltage comparator and two NPN transistors to give a constant current and also to reduce leakages across the photodiode. The switching frequency of the interface was 20 MHz, which was high enough for detecting the edge of each stripe. Thus, the rotation of the follower led to a 'pulse train' being generated from the output of the interface. This 'pulse train' was then fed to a TTL 'EXCLUSIVE OR' gate.

The variation of the duration of each pulse represented the variation of the instantaneous rotational frequency of the follower. Software was developed to convert the 'pulse train' into the instantaneous rotational frequency of the follower as a function of cam angle.

In order that the film and the infra-red emitter and detector could be installed into the right position, the valve spring and the valve guide were moved up by 33 mm (see Fig. 12.1). A slot was cut on the valve stem to hold the spring collar and subsequently the valve spring. It is believed that this minor modification will not affect the kinematics and dynamics of the valve train.

12.5 Locating the cam position

Locating the cam position is important to the later correlation between theoretical results and experimental measurements and also the average values in one cam cycle, such as the friction, the power loss, and the follower rotational frequency.

In the current test the cam position was located through a pulse generated with a slotted switch, which in turn was triggered by a one-hole disk mounted on the ebonite connector. The trigger was set on the cam base circle 180 degrees ahead of the point of the maximum cam lift. Instead of feeding the impulse signal to the third channel of the ADC as in the previous test [1], it was fed to a monostable Schmitt circuit, which reshaped the pulse generated by the slotted switch into an extremely narrow square, the width of which was only 25 μS. Then this narrow pulse was sent to the input end of the 'EXCLUSIVE OR' gate together with the 'pulse train' output from the monitoring system for the follower rotational frequency. The output from the gate was a series of pulses which related to the variation of the follower rotational frequency and a narrow pulse in each cam cycle and was sent to channel '0' of the ADC. This pulse is called the 'locating pulse'. The locating pulse synchronized to the hole on the trigger disk and the duration of the pulse was expended to 40 μS to guarantee at least three data points to be sampled (with 100 kHz sampling rate). The details of how to identify the locating pulse from the 'pulse train' and to determine the cam position are given in Section 15.4.

Since the maximum sampling rate was fixed at 200 kHz, the maximum sampling rate for each channel was dependent upon the number of channels N to be sampled and was given by 200/N kHz. The current arrangement to the cam position trigger signal saved one channel, thus the number of the total ADC channels required in the test was only two. Therefore, the maximum data acquisition rate was 100 kHz for each channel.

12.6 Sampling the torque of the camshaft

Basically the construction of the torque measurement system adopted in the present experimental study was the same as the one used in the previous tests [1]. The only modification made on the system was that the supply voltage to the strain gauge of the torque transducer was increased from 5 to 10 V. Thus, the sensitivity of the torque transducer was doubled. The torque signal was amplified and sent to channel one of the ADC.

12.7 References

1. **Zhu, G.** (1988) A theoretical and experimental study of the tribology of a cam and follower. PhD thesis, Mechanical Engineering Department, Leeds University, UK, May.
2. **Dowson, D., Taylor, C. M.,** and **Zhu, G.** (1989) An experimental study of the tribology of a cam and flat-faced follower. 2nd International Conference on the *Combustion Engine – Reduction in Friction and Wear*, pp. 97–105.

Chapter 13

Calibration of the Instrumentation and Commissioning Tests

13.1 Introduction

This chapter consists of two parts: calibration and commissioning tests. The calibration was conducted on measuring systems for the torque and the follower rotational frequency. Since the equipment used for monitoring the temperature was the same as that in the previous experimental study [1, 2] it was unnecessary to calibrate it again.

Commissioning tests are very important for obtaining consistent and repeatable results, especially with newly designed test apparatus and new instrumentation. Some problems with both mechanical construction and electronic instrumentation will be identified in commissioning tests. The current commissioning tests are concentrated on the monitoring system for follower rotation.

In the present calibration and commissioning tests, and also in the later normal experimental measurements, the parameters adopted for driving the data acquisition were kept the same. Two channels (channel 0 and 1 of the ADC) were used for sampling the torque signal and the follower rotational frequency signal respectively. The data acquisition rate was 100 kHz for each channel. The total number of data points to be sampled, therefore, was 100 000 for each channel. So, the sampling duration for each set of data was one second. This duration represented about nine cam cycles if the camshaft rotated at 550 r/min and about 34 cycles for 2050 r/min.

13.2 Calibration of the instrumentation

13.2.1 The calibration to the torque measurement system

Since the supply voltage to the strain gauge of the torque transducer was increased from 5 to about 10 V and the ADC was shifted from the VAX computer to a PC, the torque measurement system (which included the torque transducer, the voltage amplifier, the ADC and the computer were recalibrated. The system was calibrated as one 'black box' to eliminate the influence arising from each individual part of the system.

By applying a known torque to the torque transducer, the output voltage was sent to channel 1 of the ADC which was driven by a program in Microsoft C. The sampling rate and the total number of sampling points were kept the same as that to be adopted in the later tests. The sampled voltage signal was displayed on the screen and the average value of the voltage was calculated and then was written into a data file. For each applied torque such sampling and averaging operations were repeated five times. Finally, the output voltage corresponding to the applied torque to the transducer was taken to be the average value of the five averaged output voltages. The torque loaded on the transducer was increased with the increment being 0.5, 1.1, and 2.2 N m from zero to 16.7 N m in 19 steps and then decreased from 16.7 to −16.7 N m step by step. The smaller the torque applied to the transducer, the smaller was the increment adopted. A simple linear regression with 38 pairs of applied torques and output voltages was carried out. A straight line of the form

$$T = 4.5591V + 0.04871 \text{ (N m)}$$

with data points (V_1, T_1), (V_2, T_2), …, (V_{38}, T_{38}) was obtained. Figure 13.1 shows an excellent linear correlation between the applied torques and the output voltages. Moreover, the calibrations conducted after the test showed the same results.

The constant in the above linear relationship represents the zero shift of the torque measurement system. In the normal running condition this value will be changed due to the parasitic friction arising from the camshaft supporting bearings and the torque transducer. In order to measure the friction of the valve train this value should be calibrated separately under each running condition corresponding to each test condition.

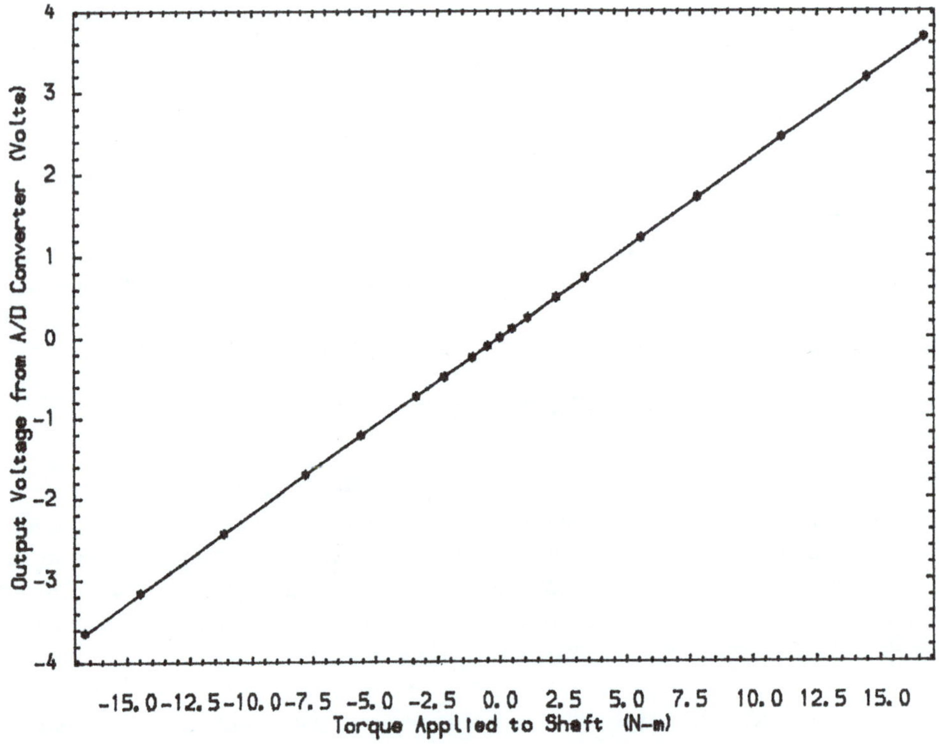

Fig. 13.1 Calibration of the torque measurement system

13.2.2 The parasitic friction of the test apparatus

In the previous tests on the valve train with a bucket follower the parasitic friction arising from the camshaft support bearings and the torque transducer was estimated from the torque signal on the cam base circle. Since there was no contact between the cam and the follower on the cam base circle due to a 0.4 mm clearance, this was a reasonable way to evaluate the parasitic friction. For the current test, however, the follower contacted with the cam in the whole cam cycle, and a post-test calibration was performed by removing the top of the test cell and running the apparatus under the same operating condition corresponding to each test run. A data acquisition program would automatically drive the ADC and the computer to carry out sampling data, calculating the average value and writing the average value into a data file at an equal interval within the whole calibrating period. The parasitic frictions in each normal test run could be estimated from the calibrated values corresponding to the equivalent moment when the experimental data were sampled. By doing so, the error caused by the thermal drift of the instrumentation and the torque transducer could also be eliminated from the experimental measurements. Therefore, the measured torque was given by

$$T = 4.5591 \times (V - V_p) \text{ (N m)}$$

where V is the voltage measured in the normal test run and V_p is the voltage sampled from the corresponding calibration run at the equivalent moment when V was sampled.

13.2.3 The calibration for monitoring the follower rotation

The calibration for the measurement of the follower rotational frequency was conducted by fixing the follower and the film on a mini-electrical motor rotating at a constant frequency. The rig for the calibration is shown in Fig. 13.2. The rotational frequency of the motor was variable and was monitored by a tachometer. The 'pulse train' output from the 'EXCLUSIVE OR' gate was fed to channel 0 of the ADC and was converted into the instantaneous rotational frequency of the follower as a function of time. This measured rotational frequency of the follower was then compared with the tachometer readings. Ideally, the measured instantaneous rotational frequency should show a straight line relationship within the sampling period. As a matter of fact the measured instantaneous rotational frequency fluctuated around the value indicated by the tachometer.

Further analysis showed that the amplitude of the fluctuation (peak to peak) was proportional to the follower rotational frequency and the sampling interval. The percentage of the deviation of the measured rotational frequency of the follower from the nominal rotational frequency could be approximately estimated by

$$\Delta N \leq \pm 3.0 \times \Delta t \times N\%$$

where Δt is the sampling interval in seconds and N is the nominal rotational frequency of the follower in r/min. In the present test two ADC channels were employed to sample the data, thus the maximum data sampling rate for each channel was 100 kHz, which was equivalent to the sampling interval being 10×10^{-6} seconds. For example, if the nominal rotational frequency is 1000 r/min the percentage deviation will be ±3.0 per cent.

Fig. 13.2 The apparatus for the calibration of the system for measuring the rotational frequency of the follower

13.3 Commissioning tests

As some modifications had been made to the test apparatus and the instrumentation, it was necessary to assess their performance capability for the operating conditions envisaged. The commissioning tests described in this section were to verify the reliability of the monitoring system for the follower rotation and the dynamic characteristics of the camshaft system. At the same time the software developed for the data acquisition and the data processing was also to be tested.

13.3.1 Improving the signal for the follower rotation

At the beginning of the commissioning test the results showed that a lot of high spikes were superimposed on the signals of follower rotational frequency. These spikes represented very narrow pulses which were generated in the system due to random interruptions to the infra-red beam. Further investigation revealed that the film on the top of the tappet was contaminated by the lubricant leaking out from the hydraulic lash adjuster. The contaminated film weakened the infra-red light received by the receiver and consequently the output current from the receiver was reduced to a level close to the critical point. Thus any slight variation within the system would cause the reverse of the output signal from the precision photodiode comparator. By moving the infra-red emitter and the receiver closer to each other, the attenuation of the infra-red light was reduced. Consequently, the interruption due to contamination of the film was eliminated.

Unfortunately, after the apparatus warmed up, spikes appeared again. At first this was thought to be due to instability of the electronic circuit, which introduced extra interruption into the measurement system. But, by moving the measurement system off the test apparatus and installing it on the rig for calibration, the spikes never appeared, no matter how long it had run, even by heating the instrument and the sensors with a hairdryer. Thus, it was clear that these spikes were not caused by a fault of the instrumentation.

Finally, it was realized that thermal expansion caused the problem. The top of the test cell was fastened on to the test cell by ten 5.5 mm diameter, 6.8 cm long screws before the test run. As the test cell reached its thermostatic condition the screws expanded, the contact between the top and the cell loosened, and consequently the up and down movement of the follower caused an extra movement of the top relative to the test cell. As the infra-red emitter and the receiver were mounted on the top of the test cell while the movement of the follower was independent of the top of the test cell, an extra movement between the film and the receiver was created. It was this extra relative movement that caused the interruption of the normal operation of the monitor system generating a series of high spikes in the measured signals. By adopting elastic washers and fastening the screws again when the test cell reached its thermostatic condition, the measured signal was greatly improved, although some spikes remained. This was due to black spots inherited on transparent stripes of the film, which was contaminated when it was developed. These remaining spikes can be eliminated by post-experiment data processing. Details will be given in Chapter 15.

13.3.2 Tests at different camshaft rotational frequencies

As described in reference [1], under certain camshaft rotational frequencies severe torsional vibration would be induced by the rotational frequency and its higher order frequency components. In order to reduce the influence of such vibrations on the torque signal, the camshaft should rotate with the frequency whose higher order frequency components are furthest away from the natural frequency of the torque transducer, which was about 165 Hz [1].

Since the rotational frequency was one of the experimental variables, a series of commissioning tests at different frequencies was carried out. It was determined that the current tests would be conducted at the camshaft rotational frequencies of 550, 850, 1150, 1450, 1750, and 2050 r/min respectively.

13.4 References

1. **Zhu, G.** (1988) A theoretical and experimental study of the tribology of a cam and follower. PhD thesis, Mechanical Engineering Department, Leeds University, UK, May.
2. **Dowson, D., Taylor, C. M.,** and **Zhu, G.** (1989) An experimental study of the tribology of a cam and flat-faced follower. 2nd International Conference on the *Combustion Engine – Reduction in Friction and Wear*, pp. 97–105.

Chapter 14

The Test Procedure

14.1 Introduction

In the present experimental study to validate the theoretical model and also to investigate the tribological performance of the valve train, the camshaft rotational frequency is one of the major test variables. In addition to the tests at different camshaft rotational frequencies in a loop test procedure as conducted in the previous experimental study [1], a duration test procedure was also adopted to confirm the repeatability and the consistency of experimental measurements.

14.2 The running-in of the valve train

In the present experimental study only one cam and tappet pair was employed throughout the whole tests. The running-in of the valve train was conducted for about 50 hours to stabilize the test apparatus. The bulk temperature was set to be 70 °C. First, the apparatus was run at a medium camshaft rotational frequency (1150 r/min) for eight hours and then shut down overnight and the lubricant was drained. Oil pipes were cleaned with compressed air and the sump of the thermostatic circulator was flushed with INHIBISOL solvent and then was recharged with fresh lubricant. Subsequently, the running-in was continued with five further procedures with the camshaft rotating at 550, 850, 1450, 1750, and 2050 r/min respectively. For each frequency the apparatus was run for eight hours at the same bulk temperature. At the end of the running-in, the lubricant was changed after the lubricant supply system had been cleaned. In the whole running-in period the torque signal and the follower rotational frequency signal were also monitored to verify the stability of the test apparatus and the instrumentation. The normal test procedure was started after the running-in.

14.3 The loop test

The term 'loop test' adopted in this report represents a test procedure to measure the experimental variables of the valve train with the camshaft rotational frequency being varied step by step with the same interval.

The instrumentation was switched on 30 minutes before starting the loop test. The apparatus was run at 1150 r/min for two hours to enable it to reach a stable condition. In the test loop, the camshaft rotational frequency was increased in five equal steps from 550 up to 2050 r/min, and then reduced from 2050 down to 550 r/min with the same steps every 15 minutes. At the end of each 15 minutes, the torque signal and the follower rotational frequency signal were sampled and stored in a data file on the hard disk. Then the camshaft rotational frequency was set to the next test frequency and run for 15 minutes. In the present test two test loops were conducted, which meant that four sets of data were sampled for each camshaft rotational frequency.

By subtracting the parasitic friction, which was calibrated by the loop calibration detailed in the previous chapter, the friction torque of the valve train can be obtained. The average friction torque of the valve train measured in the two test loops is plotted in Fig. 14.1 as a function of the camshaft rotational frequency. The results show very good consistency. A curve fit to these data points was made and is represented in the same figure by a solid line.

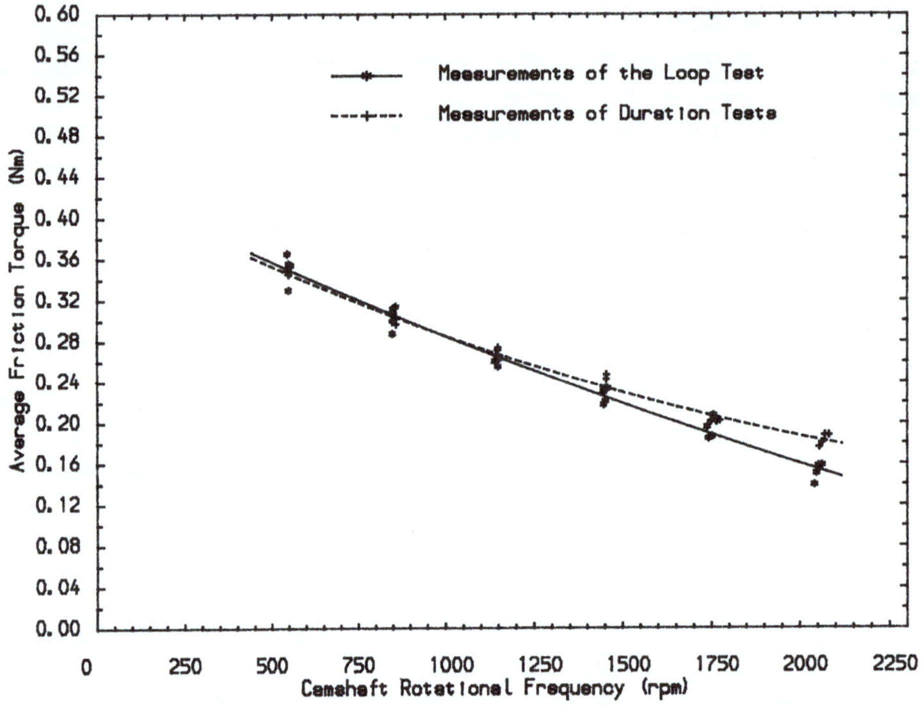

Fig. 14.1 The comparison of the average friction torque of the valve train measured in the loop test and in the duration tests

14.4 The duration test

The term 'duration test' adopted in this report represents an eight hour test procedure to measure the friction torque of the valve train at a constant camshaft rotational frequency.

In order to verify the repeatability of the experimental measurements six duration tests were performed at the camshaft rotational frequencies of 550, 850, 1150, 1450, 1750, and 2050 r/min respectively. The test was run continuously for eight hours at one camshaft rotational frequency. The test temperature was also set to be 70 °C. The torque signal and the follower rotational signal were sampled at half, one, two, four, six, and eight hours. Thus, for a given camshaft rotational frequency six sets of experimental data were obtained as a function of testing time. As described in the previous chapter, the parasitic frictions were calibrated by the duration calibration after the duration tests.

By subtracting the parasitic friction from the overall friction torque measured in the duration tests the average friction torque of the valve train was also presented as a function of the camshaft rotational frequency, see Fig. 14.1. In order to eliminate the influence of the thermal effects the last four samples in each duration test were adopted. A curve fit to these data points was also made and is represented in the same figure by a broken line. It can be seen that there is little difference between the results measured in the loop test and the duration test. This further evidence demonstrates the remarkable repeatability and the consistency of the experimental measurements. In addition, the average friction torque of the valve train was presented as a function of testing time in Fig. 14.2. It shows the friction torque reaches its stable value after only half an hour test run.

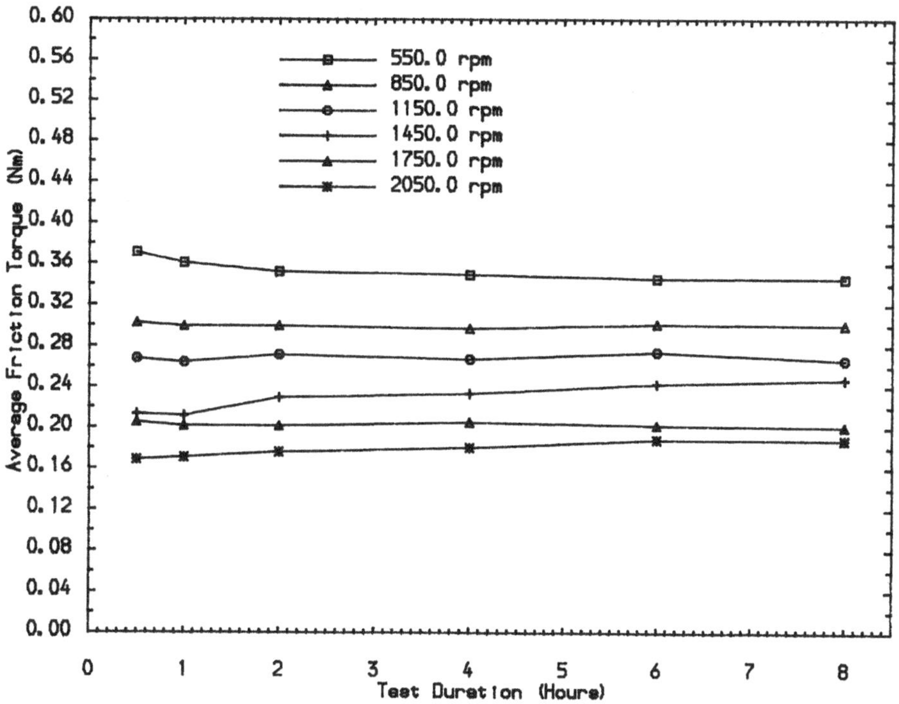

Fig. 14.2 The variation of the average friction torque of the valve train in duration tests at different camshaft rotational frequencies

14.5 Tests at different bulk temperatures

In order to simulate the operating condition of the valve train in a firing engine, the bulk test temperature was increased from 70 to 95 °C (temperature in the test cell). This is the highest temperature which can be reached in the present test apparatus, because the highest working temperature for the infra-red emitter and the receiver is about 100 °C.

As indicated in the previous section, good consistency had been achieved between the measurements of the loop test and the duration tests. Thus only a loop test was conducted on the tests for different camshaft rotational frequencies at the bulk temperature of 95 °C. Except for the interval between two test frequencies being increased to 30 minutes to achieve thermostatic conditions, other factors remained the same as for the loop test procedure described in the previous section.

14.6 References

1. **Zhu, G.** (1988) A theoretical and experimental study of the tribology of a cam and follower. PhD thesis, Mechanical Engineering Department, Leeds University, UK, May.

Chapter 15

The Data Processing

15.1 Introduction

As introduced in Chapter 12, all experimental signals were digitalized and transferred to the AMDAHL computer. The data processing was performed on the computer with a software coded in FORTRAN.

A sophisticated data processing technique had been developed [1] for the processing of the friction torque of the valve train. This technique was adopted in the present experimental study for processing the torque signal. In addition, a new technique for evaluating the instantaneous rotational frequency of the follower was developed. In order to correlate the instantaneous follower rotational frequency with the cam angle, a special function was added to the software to identify the cam position.

15.2 The torque on the camshaft

The technique for processing the torque signal developed in the previous experimental study on a valve train having a four-power polynomial cam and a flat-faced bucket follower has been detailed in reference [1]. The data processing software included a spectrum analyser, an eighth-order Butterworth low-pass digital filter, and a dynamic simulating package for the torsional vibration of the camshaft. As the mechanical construction of the camshaft driving system remained unchanged, the technique developed in the previous experimental study and related parameters such as the natural frequency of the torque transducer and the damping coefficient of the camshaft driving system were still valid for the present experimental study. This technique was, therefore, also employed in the current experimental study of the valve train having a taper cam offset to a domed follower with a hydraulic lash adjuster.

15.3 The rotational frequency of the follower

As indicated in Chapter 12, the rotational frequency of the follower was represented by a 'pulse train' characterized by a series of voltage pulses. The duration of each pulse varied in accordance with the instantaneous rotational frequency of the follower. The conversion from the 'pulse train' to the follower rotational frequency was performed by a software.

By setting a voltage level between 0 and 4 Volt, say 1 Volt, the rising or the falling edge of each pulse could be identified. The falling edge of a pulse indicated the moment when the front edge of a black stripe (or the back edge of a transparent stripe) crossed the infra-red beam. Similarly, the rising edge of a pulse represented the moment when the back edge of a black stripe (or the front edge of a transparent stripe) crossed the infra-red beam. Thus the time required for the follower to rotate one degree can be calculated out from the duration between the two front edges of two adjacent pulses. With known angular displacements θ_i and the time t_i, the rotational frequency of the follower can be evaluated by numerical differentiation. In the present analysis a three-point differential form was adopted. So, at the moment of t_i, the follower rotational frequency Ω_i can be represented by the following equation:

$$\Omega_i = \frac{d\theta}{dt} \approx \frac{(\theta_{i+1}-\theta_i)\,(t_i-t_{i-1})^2+(\theta_i-\theta_{i-1})\,(t_{i+1}-t_i)^2}{(t_{i+1}-t_i)\,(t_i-t_{i-1})\,(t_{i+1}-t_{i-1})^2}$$

It should be pointed out that the rotational frequency given by the above equation represents an average value over the period of one degree angular displacement of the follower.

It has been identified through commissioning tests (see Chapter 13) that black spots inherited on the transparent stripes caused extra interruptions which generate very narrow pulses in the normal 'pulse train'. The consequence was some high spikes superimposed on the processed follower rotational frequency signal. Since the spikes were several times higher than the maximum instantaneous rotational frequency of the follower, these spikes were eliminated by a procedure which can identify the narrow pulses responsible for the spikes.

15.4 The camshaft trigger signal

In order to correlate the instantaneous follower rotational frequency to the cam angle the cam position must be located in association with the 'pulse train'. This was achieved by generating a camshaft trigger signal within the 'pulse train'.

The camshaft trigger signal generated a 25 μs duration locating pulse synchronized to a hole on the trigger disk, which was 180 degrees ahead of the cam nose. This pulse was then mixed into the 'pulse train' through an 'EXCLUSIVE OR' gate to save one ADC channel and was extended to a 40 μs duration pulse as described in Chapter 12. The voltage of the locating pulse remained either in high or in low condition which is opposite to the voltage of the 'pulse train'. This means if the 'pulse train' was at high voltage (4 V) a low voltage locating pulse (0 V) was produced, in reverse a high-voltage locating pulse (4 V) was generated.

In order to identify the locating pulse from the 'pulse train', a software has been developed. The locating pulse can be identified by its two major characteristics. First, it is a narrow pulse with its duration being equal to or less than 40 μs. This duration is equivalent to the duration of a pulse generated by a follower rotating at 2080 r/min which was at least three times higher than the normal peak value of the follower rotational frequency. Second, it appeared only once in one cam cycle which was related to the camshaft rotational frequency. Thus the locating pulse could be identified by picking up a pulse fitting both conditions from the 'pulse train'. The front edge of the pulse was taken as the cam position of 180 degrees ahead of the cam nose.

In case the signal is triggered at the moment of either the rising or the falling edge of a pulse in the 'pulse train' by coincidence, instead of producing a locating pulse in the 'pulse train', the locating pulse will be confused with the edge of the pulse whose duration is reduced or increased by 40 μs. This is called the case of a missing locating pulse. If it happens, the locating pulse cannot be simply identified in the way described above. Fortunately, further analysis indicated that the influence of a missing locating pulse on the accuracy of the experimental measurements can be neglected. First of all, the locating pulse appears on the cam base circle where the follower rotational frequency is low (less than 100 r/min). A 40 μs difference of the duration of a pulse has little effect on the final results. Second, the case of a missing locating pulse rarely happens. By displaying the follower rotational frequency signal on the screen after each data sampling operation it can be found if the locating pulse has been missed. If it has, the data acquisition operation should be repeated. This will further reduce the chance of missing a locating pulse. Finally, even if one locating pulse is missed in the sampled data file, the program is designed to recover the pulse by inserting a locating pulse in the middle of two known locating pulses. In the same way, with the assumption of the camshaft rotating with a constant frequency, if more locating pulses are missed they can be recovered by relocating the pulses with equal intervals between the two sampled locating pulses.

15.5 References

1. **Zhu, G.** (1988) A theoretical and experimental study of the tribology of a cam and follower, PhD thesis, Mechanical Engineering Department, Leeds University, UK, May.

Chapter 16

Experimental Results and Discussions

16.1 Introduction

In this chapter, the experimental results for the friction torque of the valve train and the follower rotational frequency at different camshaft rotational frequencies and test temperatures will be presented according to the measurements from the loop tests. A discussion of the experimental results and surface inspection is also undertaken. The main conclusions which can be reached from the experimental measurements are given at the end of the chapter.

16.2 Experimental variables and the test condition

The experimental variables in the present test are the camshaft rotational frequency and the bulk temperature. Considering the normal engine rotational frequency and the torsional vibration of the torque transducer, the tests were conducted with camshaft rotational frequencies of 550, 850, 1150, 1450, 1750, and 2050 r/min. Each test was conducted at bulk temperatures of 70 and 95 °C.

The other parameters relating to the test condition are listed in Table 16.1 and are kept the same throughout the experimental study.

Table 16.1 The constant parameters of the test

Parameter	*Magnitude*	*Unit*
Radius of cam base circle	18.0	mm
Cam width	11.0	mm
Cam taper	0.0573	degree
Cam offset	1.70	mm
Cam roughness height, σ_c	0.215	µm
Maximum cam lift	9.3	mm
Radius of curvature of domed follower	19.745	m
Follower roughness height, σ_f	0.207	µm
Lubricant	TEXACO 10W/30	
Inlet oil pressure to the lash adjuster	1.035×10^5 (15)	Pa (psi)

16.3 Experimental results and discussions

16.3.1 The friction torque and power loss of the valve train

The average values presented in this chapter are taken from several cam cycles in one second. Obviously, for different camshaft rotational frequencies the number of cycles involved in each averaging is different. The higher the camshaft rotational frequency the more the cycles being averaged. For example, if the camshaft rotates at 550 r/min, the average values are calculated from eight or nine cam cycles and if the camshaft rotates at 2050 r/min, the average values are worked out from 33 or 34 cam cycles.

The average friction torque of the valve train is presented in Fig. 16.1 as a function of the camshaft rotational frequency. A least-squares curve fit to the experimental measurements has been performed and plotted as a solid line for the test temperature of 70 °C and a broken line for 95 °C. It can be seen that the average friction torque decreases as the camshaft rotational frequency increases. The increase of the camshaft rotational frequency not only resulted in an increase in the entraining velocity but also led to a reduction of the load in the cam nose region. This dual action was of great benefit in promoting the formation of an elastohydrodynamic film between the cam and follower and subsequently the reduction of the friction.

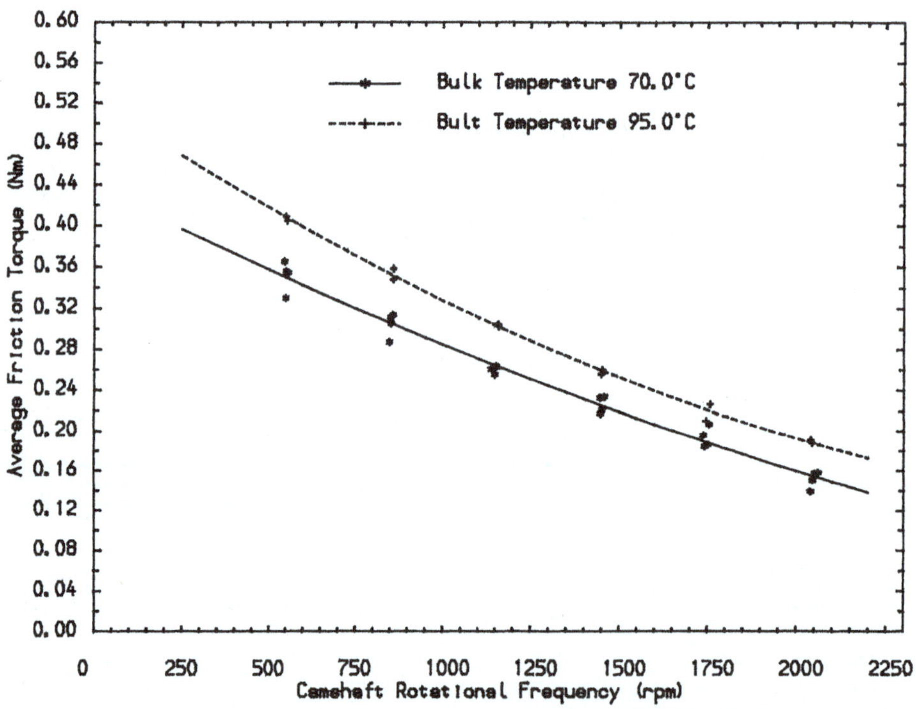

Fig. 16.1 The average friction torque of the valve train at different bulk temperatures

For the purpose of comparison the results of the tests at different bulk temperatures of 70 and 95 °C are plotted in the same figure. It can be seen that the friction torque increases as the test temperature increases. The results are consistent with the previous experimental results [1].

The product of the average friction torque with the corresponding camshaft rotational frequency gives the average power loss of the valve train. The results are plotted in Fig. 16.2 as a function of camshaft rotational frequency with the test temperatures being 70 and 95 °C. The average power loss shows an increasing trend with the increase of the camshaft rotational frequency. It is interesting to note that the power loss drops slightly as the camshaft rotational frequency rises up to 2050 r/min at the test temperature of 70 °C.

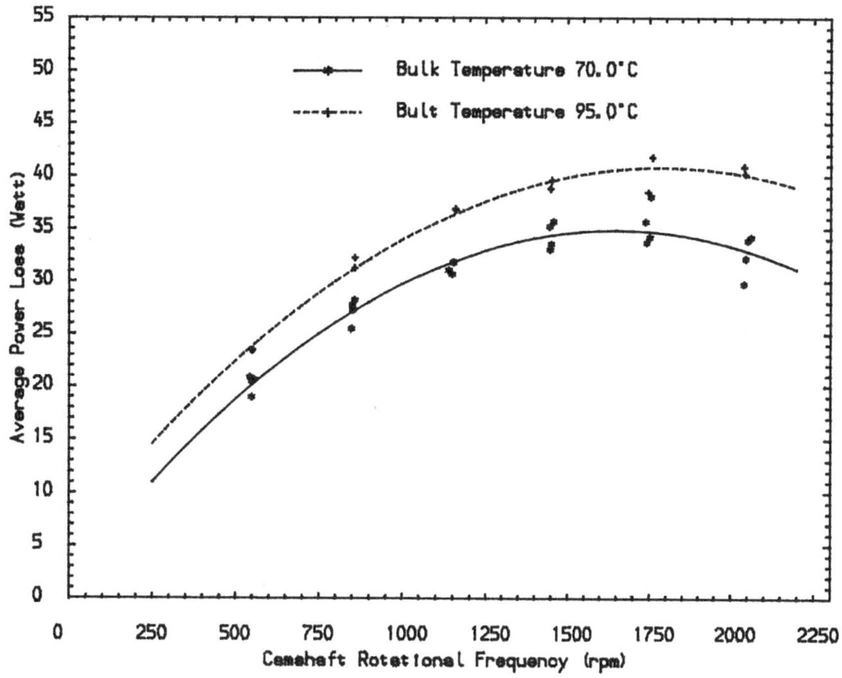

Fig. 16.2 The average power loss of the valve train at different bulk temperatures

Since the instantaneous friction torque, friction, and friction coefficient of the valve train are associated with the cam load which is estimated from the theoretical model, they will be presented in Part III, the correlation between the theory and the experiment. Figures 16.3 and 16.4 show the cyclic variation of torques of the camshaft rotating at different frequencies for two bulk temperatures. At low camshaft rotational frequency (say below 1000 r/min) the camshaft torque varies in the pattern of a sine-wave. As the camshaft rotational frequency increases a twin peak (or twin valleys) pattern is formed, which can be seen, for example, from Fig. 16.3(d), (e), and (f). The increase of the inertia of the valve train due to high camshaft rotational frequency leads to a substantial increase of the load on the cam flanks and consequently forms another peak (or valley).

16.3.2 The rotational frequency of the follower

In order to demonstrate the cyclic variation of the follower rotational frequency, the instantaneous values are plotted as a function of time in Figs 16.5 and 16.6. Each cam cycle is identified by the locating pulse (180 degree cam angle ahead of the cam nose located at 0 degrees) and is marked on the figures with short, vertical lines. As

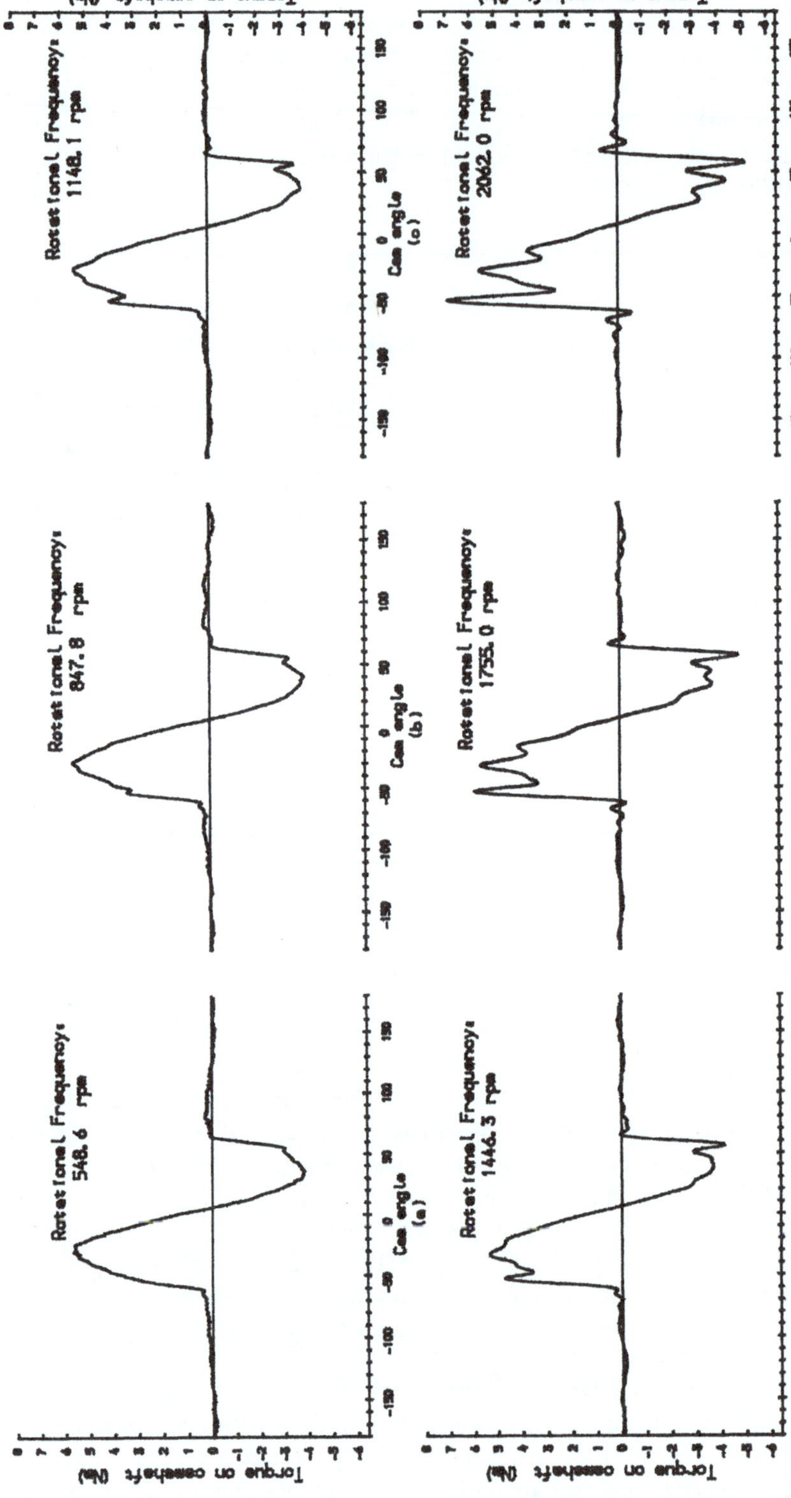

Fig. 16.3 Comparison of the instantaneous torque on the camshaft rotating at different frequencies (lubricant Texaco 10W/30, bulk temperature 70 °C)

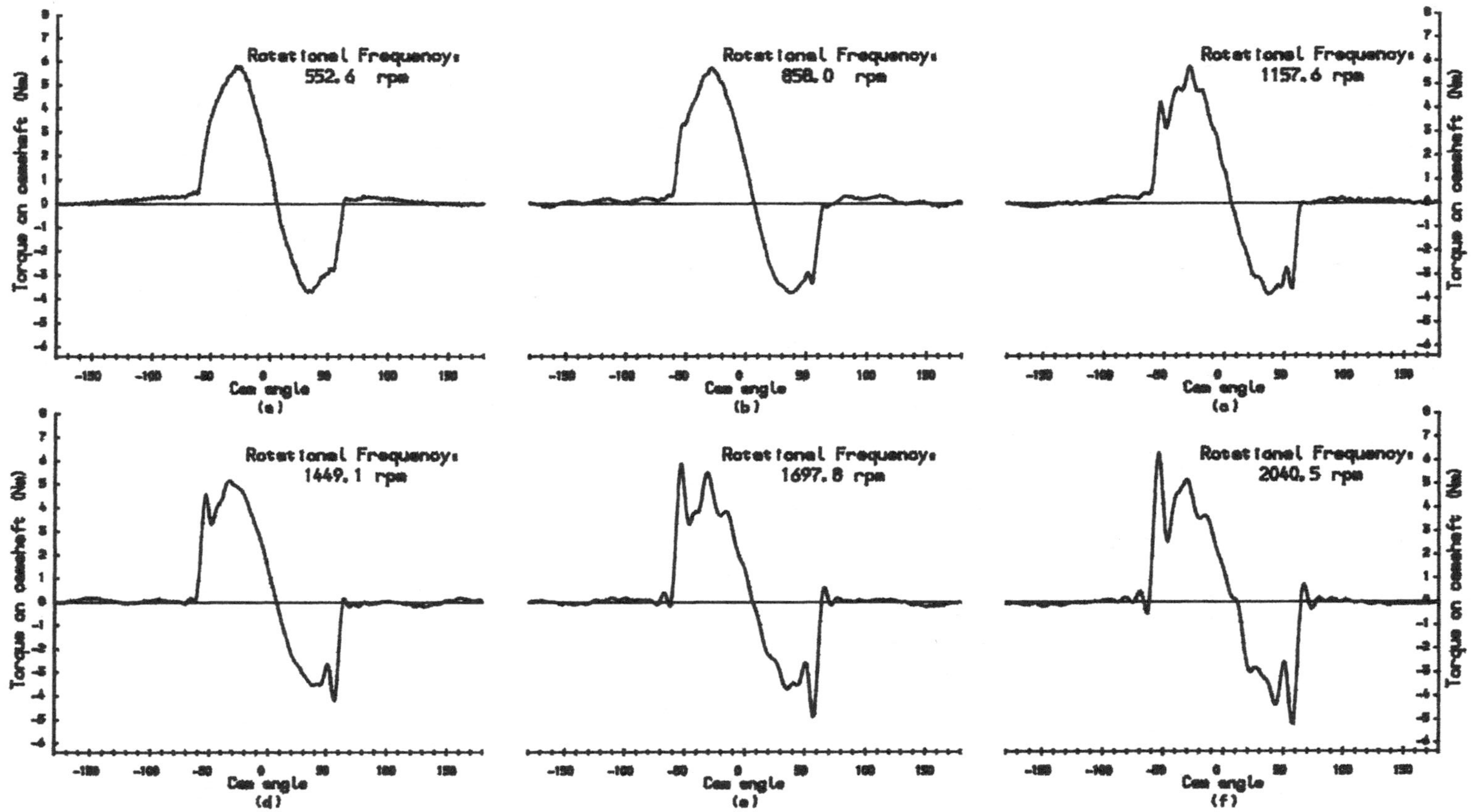

Fig. 16.4 Comparison of the instantaneous torque on the camshaft rotating at different frequencies (lubricant Texaco 10W/30, bulk temperature 95 °C)

described in Chapter 13 the sampling duration is one second. So, the variation of the follower rotational frequency in eight full cam cycles with the camshaft rotating at 548 r/min and the bulk temperature being 70 °C is shown in Fig. 16.5. For a higher camshaft rotational frequency of 2050 r/min with the same bulk temperature, Fig. 16.6 presents the variation of the follower rotational frequency in 33 full cam cycles.

There are several interesting characteristics which should be discussed in detail.

1. The instantaneous follower rotational frequency seems to be quite irregular with the maximum follower rotational frequency fluctuating from cycle to cycle. Obviously, the average follower rotational frequency is also different from cycle to cycle. Sometimes the fluctuation may be substantial. For instance, in Fig. 16.5 the maximum average follower rotational frequency is 189 r/min, while the minimum value is only about 87 r/min. This finding is consistent with the previous experimental observations reported by Dyson and Naylor [2]. Since the follower rotation is affected by so many factors, such as the clearance between the tappet and bore, the actual contact point between the cam and follower, the tilting of the tappet, and the position of the tappet relative to the inlet oil hole, this irregularity is to be expected.

2. By taking a close look at Fig. 16.6, although the average follower rotational frequency fluctuates from cycle to cycle, this fluctuation shows a cyclic variation in the whole sampling period (one second). This reveals that the 'irregular' rotation of the follower is due to the irregularity of the mechanical construction of the valve train and not some random factor. It is, therefore, still possible to investigate the follower rotation at different operating conditions by adopting an average follower rotational frequency, if this value is averaged from a number of cam cycles. As mentioned above for different camshaft rotational frequencies the number of cycles being averaged is different. The average follower rotational frequency is presented in Fig. 16.7 as a function of the camshaft rotational frequency for bulk temperatures of 70 and 95 °C. It can be seen that the scatter of the average values for the same operating condition is much smaller (less than 20 r/min) than the averages from each individual cam cycle. For both cases, the average follower rotational frequency increases as the camshaft rotational frequency increases. In addition, increasing the test temperature also leads to an increase of the average follower rotational frequency, especially at high camshaft rotational frequency.

3. It can be seen from Fig. 16.7 that the average follower rotational frequency is only about one tenth of the camshaft rotational frequency. This indicates that for the test condition there is substantial slip between the two surfaces of the cam and the follower, although the follower is free to rotate. It is contrary to the conventional view that at the contact point the follower moves in the same direction with the same velocity as the cam. It is, therefore, expected that reduction of friction due to the rotation of the follower is limited.

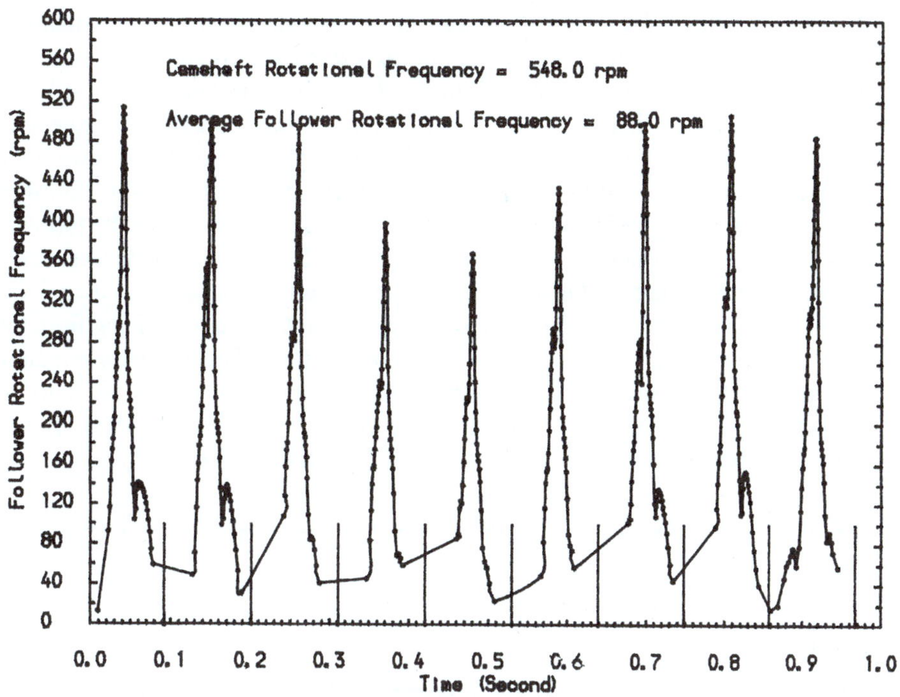

Fig. 16.5 The cyclic variation of the instantaneous follower rotational frequency (lubricant Texaco 10W/30, bulk temperature 70 °C)

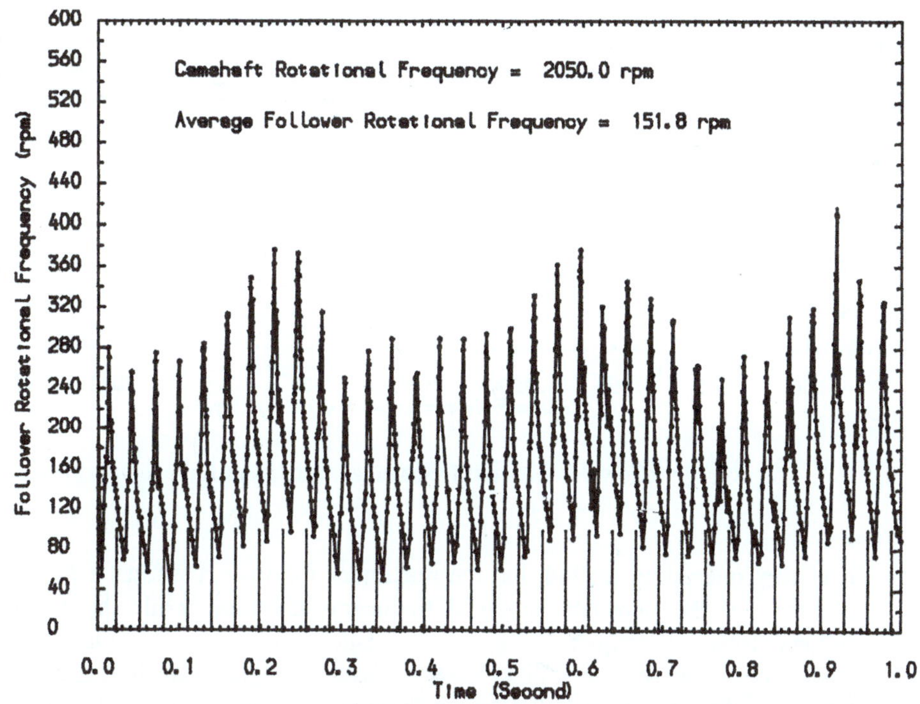

Fig. 16.6 The cyclic variation of the instantaneous follower rotational frequency (lubricant Texaco 10W/30, bulk temperature 70 °C)

4. In spite of the individual difference, the instantaneous follower rotational frequency varies with the same pattern in each cam cycle as shown in Figs 16.5 and 16.6. In order to investigate more details, the instantaneous follower rotational frequency in one cam cycle at different camshaft rotational frequencies is plotted in Figs 16.8 and 16.9 as a function of cam angle for two bulk temperatures. The figures demonstrate that the variation of the follower rotational frequency follows a similar pattern: on the cam base circle the follower rotational frequency decreases from the end of the closing ramp to the start of the opening ramp of the cam. On the opening ramp it starts increasing and then drops on the opening flank due to a substantial reduction of the friction arising from the cam/follower interface. As the contact point moves towards the cam nose region, the friction increases and consequently the rotational frequency rises to its maximum value and then decreases as the contact moves to the cam closing ramp, followed by another slight increase of the rotational frequency.

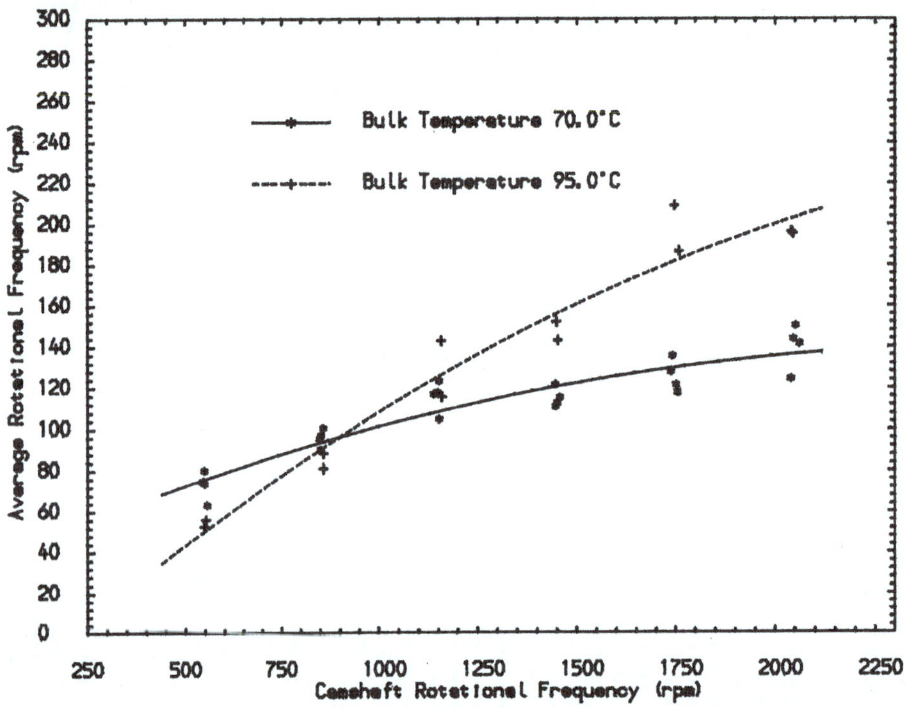

Fig. 16.7 The average follower rotational frequency at different bulk temperatures

5. It is very interesting to find that the maximum follower rotational frequency decreases as the camshaft rotational frequency increases, while the average follower rotational frequency shows an opposite trend in Fig. 16.7. On the cam base circle, however, the follower rotates faster at higher camshaft rotational frequency than it does at lower camshaft rotational frequency. In other words, the higher the camshaft rotational frequency, the more evenly the follower rotates.

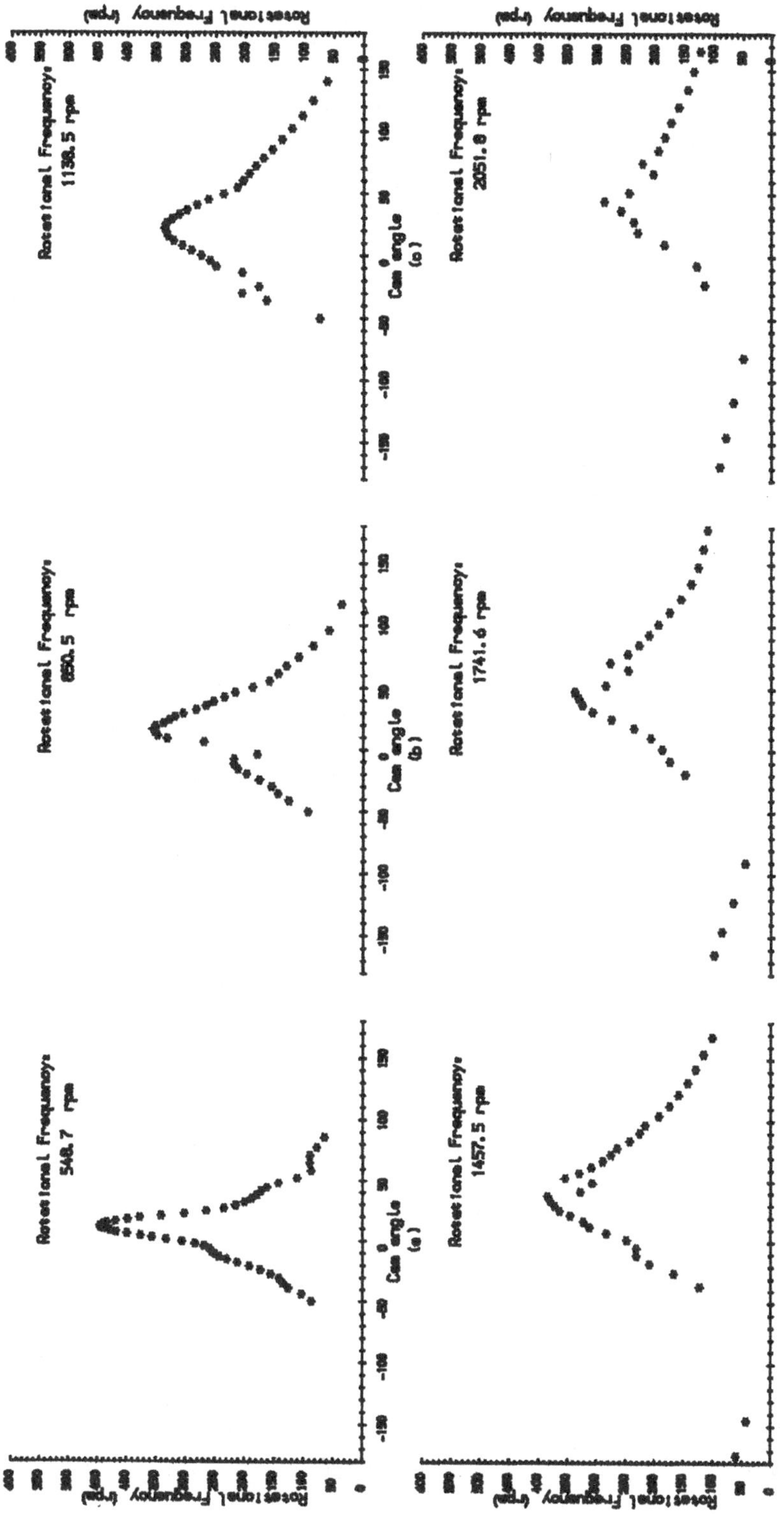

Fig. 16.8 Comparison of the instantaneous follower rotational frequency at different camshaft rotational frequencies (lubricant Texaco 10W/30, bulk temperature 70 °C)

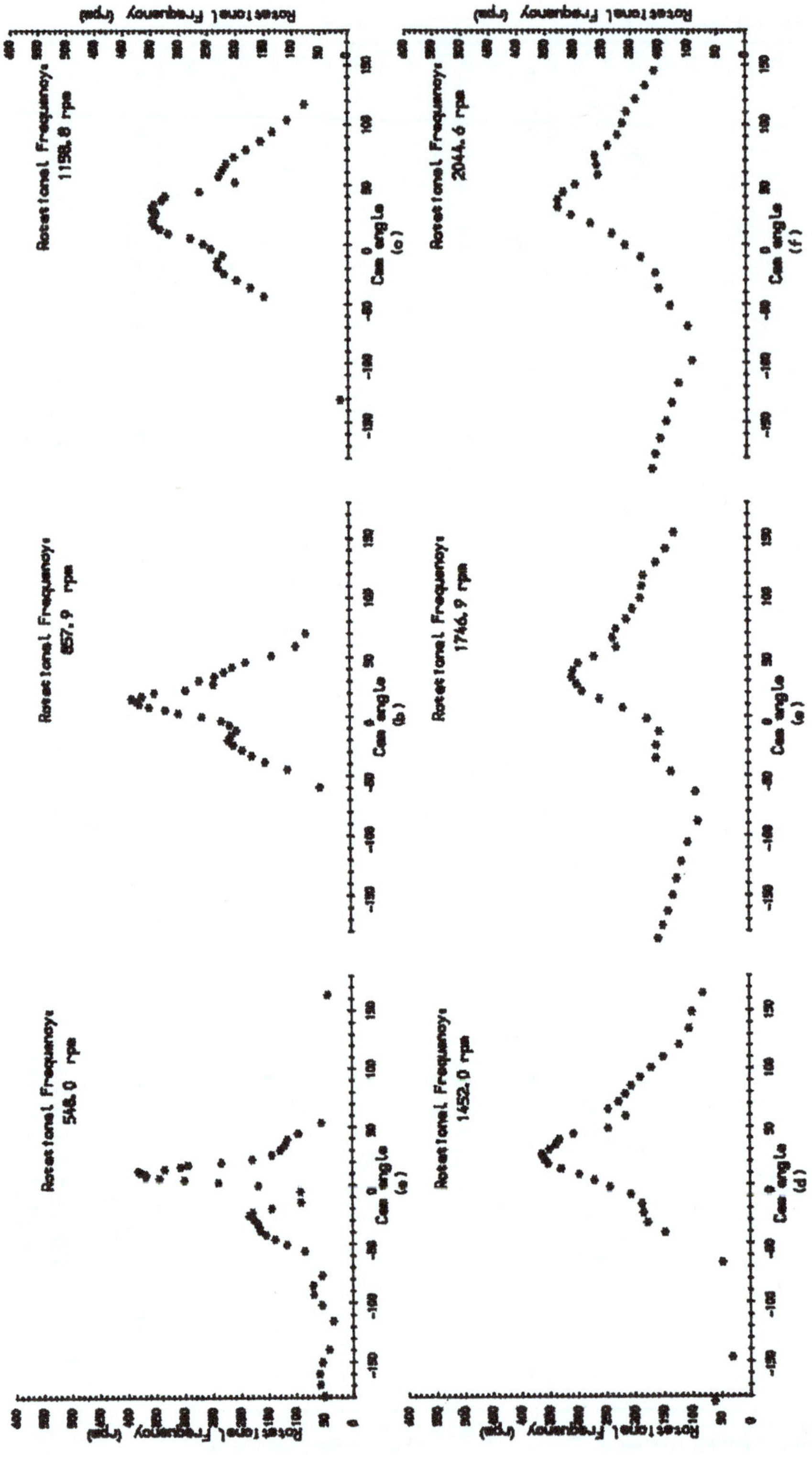

Fig. 16.9 Comparison of the instantaneous follower rotational frequency at different camshaft rotational frequencies (lubricant Texaco 10W/30, bulk temperature 95 °C)

The analysis of the dynamics of follower rotation reveals that follower rotation is dependent upon the balance of the following three factors: first, the friction arising from the interface of the cam/follower and the shifting of the contact from the centre of the follower provide a torque which causes the follower to rotate; second, the acceleration and deceleration of the follower rotation are associated with the moment of inertia of the follower and how long it is accelerated or decelerated; finally, the friction of the interfaces of the follower/bore and the plunger/valve resists follower rotation. In the case of a camshaft rotating at low frequency, the follower receives a higher driving torque (because of the high friction of the cam/follower interface) and has a longer duration to be accelerated or decelerated; the consequence is that the follower rotates slower on the cam base circle and can reach higher rotational frequency in the cam nose region than it does at high camshaft rotational frequency. The increase of the averaged follower rotational frequency along with increasing the camshaft rotational frequency is associated with the increase of the frequency of acceleration (the times of the follower being accelerated in unit time). Since the friction arising from the cam/follower interface decreases as the camshaft rotational frequency increases (see Fig. 16.1), it is expected that the average follower rotational frequency will reach a maximum value. A further increase of the camshaft rotational frequency will have little effect on increasing follower rotation and sometimes a slight decrease of the average follower rotational frequency might occur.

The average follower rotational frequency increases when the bulk temperature increases from 70 to 95 °C as demonstrated in Fig. 16.7. This is caused by the following two factors: first, the increase of the driving torque, which rotates the follower, as a result of the increase of the friction along with the rise of the bulk temperature as shown in Fig. 16.1. Second, the resistance arising from the tappet/bore interface decreases due to the increase of the clearance and the decrease of the lubricant viscosity at a higher temperature.

16.3.3 Inspection of the cam and the follower

After the test the apparatus was disassembled. The cam and the follower were inspected by eye and Talysurf machine. This inspection provides some important evidence to validate the theoretical model developed in the first part of this book and to reveal the real contact situation between an offset taper cam and a domed follower.

The general view of the follower surface after a test shows a lot of polished concentric circles. This is evidence indicating the rotation of the follower. The comparison of the Talysurf traces measured before and after the test demonstrates that the central part of the follower, a 10 mm diameter circle, is worn flat with slight concavity. This implies the possibility of the actual contact point between the cam and the follower deviating from its nominal position towards the follower centre due to the tilting of the follower. The inspection of the cam surface confirms this conjecture.

In order to identify the tested cam surface, four levels (original surface, slight, medium, and high polish) were adopted according to the degree of the surface polish. They represent the different levels of the surface contact (or the lubrication condition) in association with the cam and the follower. The cam surface is unfolded along the cam profile in Fig. 16.10 to show the distribution of the different polishing level. In the figure two terms: 'high edge' and 'low edge' are adopted to identify each edge of the taper cam. The 'high edge' represents the cam edge at the big end of the cam and is shifted away from the centre of the follower due to the cam offset and, consequently, the other cam edge, 'low edge', is shifted towards the centre of the follower.

As indicated in the figure the high polish occurs in the cam nose region in which severe asperity contact is expected due to extremely low lubricant film thickness and high load. On both cam flanks, the surface remains untouched (original surface) as a result of a healthy lubricant film. It is very interesting to note that on the cam base circle and two cam ramps all different levels are detected except for the high polish case. A two millimetre wide strip along the high edge is found to be the original surface indicating no surface contact. Within the slight polish region there is a two millimetre wide strip showing medium polish. It is believed to be the central region of the contact between the cam and follower.

It is known that the nominal shift δ_t of the first point of the contact between the cam and the follower is given by

$$\delta_t = R_f \sin \alpha'$$

According to the measurements of the Talysurf machine, the cam taper α' is about 3.4′ and the radius of curvature of the domed follower R_f is about 19.745 m. Thus, the nominal contact point deviates from the follower centre by about 19.74 mm which is far beyond the edge of the follower. Consequently, edge loading is expected according to the analysis in reference [3]. In fact, there is no contact on the high edge of the cam as described above. Moreover, it is interesting to find that the location of the strip deviates from the central plane of the cam towards the low edge by only about 1.5 mm. Considering the cam offset is 1.7 mm (see Table 16.1), it is conjectured that the actual contact between the cam and the follower on the cam base circle happens in the vicinity of the centre of the follower. There are two possibilities leading to this situation. One is the follower tilting and the other is the tilting of the camshaft, both of which compensate the cam taper. However, whatever the camshaft tilting is, the tilting of the follower always makes the contact point move towards the follower centre. This is also one of the reasons why the follower rotational frequency decreases on the cam base circle.

16.4 Conclusions

A careful and comprehensive experimental programme consisting of a series of tests at different camshaft frequencies and bulk temperatures has been conducted to investigate the tribological performance of a valve train having an offset taper cam and a domed follower, and to validate the theoretical model developed for predicting the tribological performance. Several important variables including the friction torque, power loss, follower rotational frequency, and temperatures in the valve stem guide and in the oil inlet to the hydraulic lash adjuster have been monitored in each test. A sophisticated data acquisition system and software for data processing have been developed. The surface characteristics of the cam and follower have also been inspected before and

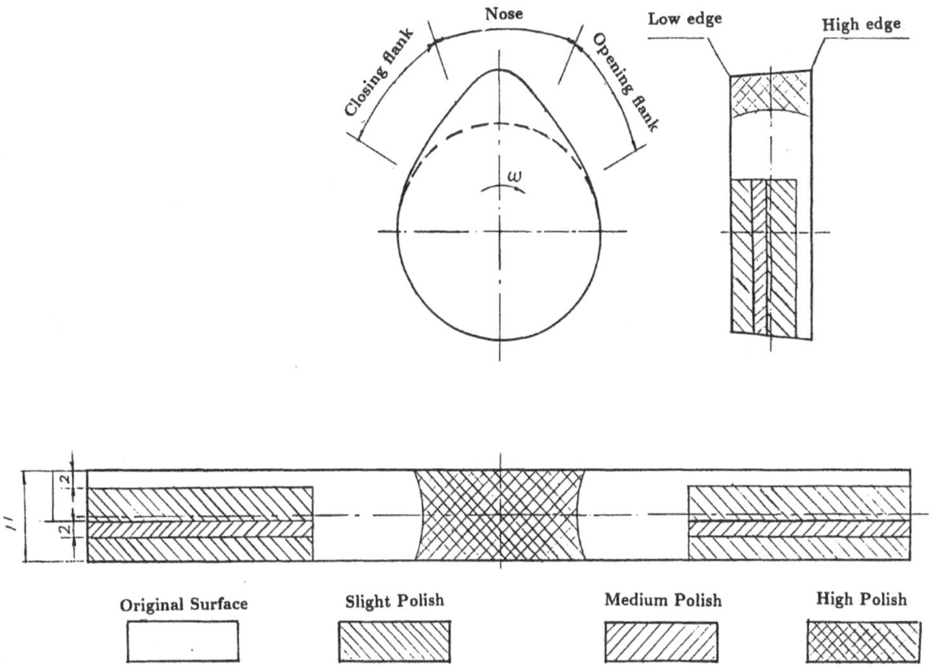

Fig. 16.10 The inspection of the cam surface after the test

after the test. The main conclusions which can be drawn from the experimental results are summarized below.

1. The excellent consistency and repeatability of the experimental measurements obtained in both loop tests and duration tests demonstrates that the test apparatus, instrumentation, test procedures, and the software for data processing developed in the current experimental study are reliable and can provide repeatable measurements with great confidence. The experimental facilities and technology developed in the current experimental study can readily be used to investigate the tribological performance of various configurations of direct-acting valve trains.

2. The averaged friction torque of the valve train decreases as the camshaft rotational frequency increases. The increase of the bulk temperature leads to the increase of the average friction torque of the valve train.

3. The average follower rotational frequency increases with increase of the camshaft rotational frequency. The increase of the bulk temperature results in an increase of the average rotational frequency of the follower.

4. The maximum instantaneous rotational frequency of the follower decreases as the camshaft rotational frequency increases, while the rotational frequency of the follower on the cam base circle increases.

5. The inspection of the cam and the follower surface before and after the tests indicates that the actual contacting centre on the base circle is in the vicinity of the centre of the follower.

16.5 References

1. **Zhu, G.** (1988) A theoretical and experimental study of the tribology of a cam and follower. PhD thesis, Mechanical Engineering Department, Leeds University, UK, May.
2. **Dyson, D.** and **Naylor, H.** (1960–61) Application of the flash temperature concept to cam and tappet wear problems. *Proc. Instn Mech. Engrs (Automobile Division),* No. 8, 225–280.

Part III – Correlation of Theory and Experiments

Chapter 17

Introduction to Correlation of Theory and Experiments

In the final part of this book, a detailed introduction to the correlation of experimental measurements and theoretical predictions of the tribological performance of the valve train is given. This includes a brief introduction to the experimental study and the theoretical model developed for predicting the tribological performance of the valve train. As has been described in the first part of the book, an advanced theoretical model has been developed to predict the tribological performance of an automobile valve train having a taper cam offset to a domed follower with hydraulic lash adjuster. The development of the model is to provide a useful tool for the tribological design of the engine valve train. Several important characteristics of the valve train can be predicted by this model. This includes the information in association with the cam/follower interface:

(a) the kinematics and dynamics;
(b) the maximum Hertzian stress;
(c) the central and minimum lubricant film thickness;
(d) the load and friction due to asperity contact; and
(e) the friction and power loss.

In addition the model can estimate the instantaneous and average rotational frequency of the follower. It is believed that this is the first time that the follower rotation has been simulated by a theoretical model. Furthermore, the performance of the hydraulic lash adjuster can also be predicted by the current theoretical model. The friction arising from the other interfaces of valve train components, such as tappet/bore and valve stem/guide interfaces can also be estimated by the model.

A careful and comprehensive experimental programme consisting of a series of tests at different camshaft rotational frequencies and different bulk temperatures has been undertaken and has been detailed in the second part of the report. This is to validate the theoretical model developed for predicting the tribological performance of the valve train and to investigate its performance. Several important variables including the friction torque, the power loss, the follower rotational frequency, and the temperatures of the lubricant supply, inside the test cell and in the valve guide have been monitored in each test. Some sophisticated techniques for data acquisition and data processing including a PC-based data acquisition system and a follower rotational frequency monitoring system have been developed.

The experimental data presented in the second part of the book for torque, average friction torque, average power loss, and average rotational frequency of the follower as a function of camshaft rotational frequency, the cyclic variation of the friction torque, and the follower rotational frequency clearly demonstrate the tribological performance of a valve train having an offset taper cam and a domed follower with hydraulic lash adjuster. The completion of the comprehensive experimental programme has provided a vast amount of reliable data. This enables the validity of the theoretical model to be examined.

In this final part of the book, the experimental measurements have been correlated with the theoretical predictions which simulate the test conditions of the valve train. In order that the correlation between the theoretical predictions and the experimental measurements can be made, the input data to the theoretical model was taken from the test apparatus to simulate the test conditions. The agreement between the experimental measurements and theoretical predictions is excellent. This gives great confidence in dealing with novel and alternative designs using the theoretical model.

Chapter 18

The Experimental Evidence

A comprehensive experimental programme consisting of a series of tests at different camshaft rotational frequencies and different bulk temperatures has been undertaken. The aim of the experimental study was to achieve the following three objectives:

1. In order that the tribological performance of different configurations of a direct-acting valve train can be estimated, it is necessary to create a stable test apparatus and a reliable data acquisition system, to set up a standard test routine, and to develop a sophisticated technique for data processing.

2. The experimental study was to investigate the tribological performance of a specific direct-acting valve train which has an offset taper cam and a domed follower with a hydraulic lash adjuster. The experimental study was focused on the intake valve train of the 1.8/2.0 litre high-output performance ZETEC engine. Further experimental study focused on the influence of some important variables, such as the camshaft rotational frequency and the bulk temperature, on the tribological performance of the valve train.

3. As has been described in Part I, an advanced theoretical model has been developed to predict the tribological performance of an automobile valve train having a taper cam offset to a domed follower with hydraulic lash adjuster. It is, therefore, necessary to validate the model with experimental evidence which includes the measurement of the rotational frequency of the follower and the friction torque of the valve train.

To achieve these objectives an optical transmission technique has been adopted to monitor the follower rotation. The top of the test cell has been redesigned to accommodate the hydraulic lash adjuster and an infra-red emitter and an infra-red receiver. A PC-based data acquisition system has been developed to provide a much higher sampling rate and longer sampling duration, with much lower noise and distortion of the signal.

Several important variables including the friction torque, the power loss, the follower rotational frequency, and the temperatures of the lubricant supply, inside the test cell and in the valve guide have been monitored in each test. Some sophisticated techniques for data processing have also been developed.

Two test procedures, the loop test and the duration test, have been adopted to examine the repeatability and consistency of the experimental measurement and to investigate the tribological performance of the valve train at different camshaft rotational frequencies. The results show that both test procedures produce the same results for the same operating conditions. This indicates that the test apparatus, the instrumentation, the data processing, and test procedures developed in the current experimental study are very reliable and can be used for further studies on the tribological performance of a direct-acting valve train with different configurations.

The experimental variables are the camshaft rotational frequency and the bulk temperatures. The tests have been conducted with camshaft rotational frequencies of 550, 1150, 1450, 1750, and 2050 r/min at bulk temperatures of 70 and 95 °C respectively. The experimental data presented in Part II for torque, average friction torque, average power loss, and average rotational frequency of the follower as a function of camshaft rotational frequency, the cyclic variation of the camshaft torque, and the follower rotational frequency as a function of cam angle have been used to correlate with the predictions of the theoretical model.

Detailed introductions of the test apparatus, the calibration and commissioning tests, data processing, test procedure, experimental measurements, and a discussion of the results have been recorded in Part II.

Chapter 19

The Theoretical Predictions

19.1 Introduction

The experimental results presented in Part II will be compared with predictions from the theoretical model developed which has been coded on an AMDAHL 580 computer. The detailed considerations relating to this model, the numerical analysis techniques and the solution procedures have been given in Part I. It is, however, appropriate to outline the basis of the model in the following sub-sections.

19.2 Kinematics and dynamics of the valve train

The kinematic analysis of the valve train is carried out to provide information about the velocity of the point of contact relative to the cam or the follower and to estimate the velocity and the acceleration of the valve for predicting the load on the cam. At the same time some geometric parameters, such as the radius of curvature of the cam and the eccentricity of the contact point, are also evaluated.

The force associated with a cam and follower pair needs to be known from both the lubrication and stress points of view. In general, the forces associated with the operation of a cam mechanism are: the inertia force, the spring force, the dynamic deflection, and damping effect associated with the parts and the friction between the moving parts. In the theoretical analysis the assumption of a rigid mechanism was made, the load between the cam and follower being concentrated in the centre of the contact.

19.3 The cam/follower interface

Predicting the tribological performance of the cam/follower interface is one of the major objectives of the theoretical model. The model provides information on a valve train including the maximum Hertzian stress, the minimum and the central lubricant film thickness, the load and friction due to asperity contact, the friction, and power loss.

In order to reduce the likelihood of scuffing and pitting failure of the cam and follower it is important to keep the contact stress as low as possible. The maximum Hertzian stress is adopted as a design criterion in cam and follower design. In addition to this, elastohydrodynamic lubrication analysis has shown that the Hertzian pressure distribution is a very good approximation to the oil pressure distribution within the contact region of the cam and follower. It has also been shown that the friction of the junction is dominated by the shear stresses within the contact region which can also be predicted from the Hertzian theory. The contact between a taper cam and a domed follower is geometrically equivalent to an ellipsoid in contact with a plane. The Hertzian theory [1] was therefore adopted to determine the maximum Hertzian stress, the pressure distribution, and ellipticity ratio of the contact ellipse in the model.

Predicting the oil film thickness between a taper cam and domed follower is one of the main tasks for the tribological analysis. The lubricant entraining velocity is one of the major parameters for predicting the oil film thickness. Due to the rotation of the follower, however, the entraining velocities varies in both magnitude and direction at each point within the contact area. In the theoretical analysis the average entraining velocity, which was the velocity at the centre of the contact ellipse, was taken to be the lubricant entraining velocity for estimating the central and the minimum film thickness. The model enabled a quick quasi-static analysis of film thicknesses associated with elliptical contacts (a taper cam and domed follower) with the direction of the lubricant entrainment at some intermediate angle to the major axis of the Hertzian contact ellipse after Chittenden, Dowson, and Taylor [2, 3]. In addition, the model also included a consideration of surface roughness effects on the hydrodynamic load by subtracting the contribution of the asperity contact from the cam load.

Both theoretical and experimental studies have shown that the surface roughness has important influences on the tribological characteristics of the cam and follower. The asperity contact force (or the load carried by asperities) and the real area of contact can be predicted in the model after Greenwood and Tripp [4].

The friction of the junction between the cam and follower is caused by two different mechanisms: the shear of the lubricant and the asperity contact. In the present model these two parts of the friction were estimated separately and then summed with regard to the proportion of the asperity contact area. Elastohydrodynamic lubrication theory showed that the friction due to the shear of the lubricant within the conjunction of the cam and follower was dominated by the sliding contribution from the 'Hertzian' region which was assumed to be parallel and subjected to the Hertz pressure distribution. The central film thickness was taken to be the separation of the two parallel surfaces. The rheological model developed by Evans and Johnson [5] was adopted in the present analysis in which non-Newtonian effects of the lubricant were taken into consideration. The friction arising from the asperity contact was assumed due to the asperity sliding under a boundary lubrication condition [6] and was equal to the product of the friction coefficient of boundary lubrication and the asperity load.

19.4 The follower/bore interface

The experimental evidence (see reference [7] and Section 16.3) suggested that the tappet tilted as it travelled up and down in the tappet bore. As pressurized oil was fed into the gap between the follower and the bore, and also the tappet was designed to be free to rotate, the lubrication condition of the interface might be influenced by hydrodynamic, mixed, or boundary lubrication. The friction arising from the interface was therefore evaluated according to its lubrication condition. The determination of the lubrication condition was dependent upon the tilting of the follower. An equivalent model which was based upon the short bearing theory [8] had been developed to estimate the follower tilting.

19.5 The hydraulic lash adjuster

A theoretical model had been developed to predict the performance of the hydraulic lash adjuster. This was essential in considerations of the dynamics of the valve train and was also important for the tribological analysis of the valve train. In the analysis it was assumed that the hydraulic lash adjuster worked with either the check valve open or the check valve closed. The theoretical model was developed separately in respect of each condition.

19.6 The rotation of the follower

The model also enabled the rotational frequency of the follower to be predicted in the whole cam cycle. The balance of external moments acting on the follower, caused by the frictions arising from the interface of the cam/follower, follower/bore and valve/plunger, and the moment of inertia of the follower yielded to a non-linear two-order ordinary differential equation. The numerical integration of the equation gave an instantaneous rotational frequency of the follower at a cam angle. Starting with an estimated initial rotation frequency of the follower the analysis was conducted in parallel with other tribological analyses at each cam angle over several cam cycles until a convergent result was obtained.

19.7 References

1. **Hamrock, B. J.** and **Dowson, D.** (1981) *Ball Bearing Lubrication – the Elastohydrodynamics of Elliptical Contacts* (John Wiley & Son, New York).
2. **Chittenden, R. J., Dowson, D., Dunn, J. F.**, and **Taylor, C. M.** (1985) A theoretical analysis of the isothermal elastohydrodynamic lubrication of concentrated contacts I. Direction of lubricant entrainment coincident with the major axis of the Hertzian ellipse. *Proc. R. Soc. Lond.*, **A397**, 245–269.
3. **Chittenden, R. J., Dowson, D., Dunn, J. F.**, and **Taylor, C. M.** (1985) A theoretical analysis of the isothermal elastohydrodynamic lubrication of concentrated contacts II. General case, with lubricant entrainment along either principle axis of the Hertzian ellipse or at some intermediate angle. *Proc. R. Soc. Lond.*, **A397,** 271–294.
4. **Greenwood, J. A.** and **Tripp, J. H.** (1970–1971) The contact of two nominally flat rough surfaces. *Proc. Instn Mech. Engrs,* **185,** 625–633.
5. **Evans, C. R.** and **Johnson, K. L.** (1986) The rheological properties of elastohydrodynamic lubricants. *Proc. Instn Mech. Engrs, Part C,* **200** (5), 625–633.
6. **Nemlekar, P. R.** and **Cheng, H. S.** (1980) Traction in rough elastohydrodynamic contacts. In *Surface Roughness Effects in Hydrodynamic and Mixed Lubrication* (Eds M. S. Rohde and H. S. Cheng).
7. **Zhu, G.** (1988) A theoretical and experimental study of the tribology of a cam and follower. PhD thesis, Mechanical Engineering Department, Leeds University, UK, May.
8. **Pinkus, O.** and **Sternlicht, B.** (1961) *Theory of Hydrodynamic Lubrication* (McGraw-Hill Book Company, Inc.).

Chapter 20

The Comparison of Results and Discussions

20.1 Introduction

The correlation of the theoretical predictions and the experimental measurements is presented in this chapter. The average friction torque of the valve train, the instantaneous torque on the camshaft, the instantaneous friction torque of the valve train, the average rotational frequency of the follower, and the instantaneous rotational frequency of the follower predicted from the model have been compared with the experimental measurements. A discussion of the correlation of theoretical and experimental results is also given in this chapter.

20.2 The input data of the model simulating the test conditions

A sample analysis of the theoretical model developed was given in the first part of the book, which corresponded to the design case of the intake valve train of the 1.8/2.0 litre high-output performance ZETEC engine. All input data (design parameters of the valve train) required for the theoretical analysis were taken from the nominal dimensions of component drawings of the valve train of the engine provided by the Ford Motor Company. The radial clearance between two components was taken to be the mean value according to the tolerance limits. In the present theoretical analysis some parameters corresponding to the tested components are not available, such as cam profile, the clearances of components of the hydraulic lash adjuster, and some rheological parameters relating to the lubricant, and therefore these parameters are taken to be the same as the design case.

The lubricant used in the tests was TEXACO 10W/30. The viscosity–temperature characteristics had been determined with a U-tube viscometer (type BS/IP/SL) at the Institute of Tribology, University of Leeds [1]. The variation of the dynamic viscosity of the lubricant with temperature is shown in Fig. 20.1. The solid line was an exponential curve fit to measured data pairs. With known bulk temperature of the tests of the valve train being 70 and 95 °C respectively, the lubricant viscosity at atmospheric pressure was calculated from the empirical formula which was an exponential curve fit to the calibrated data pairs of the lubricant.

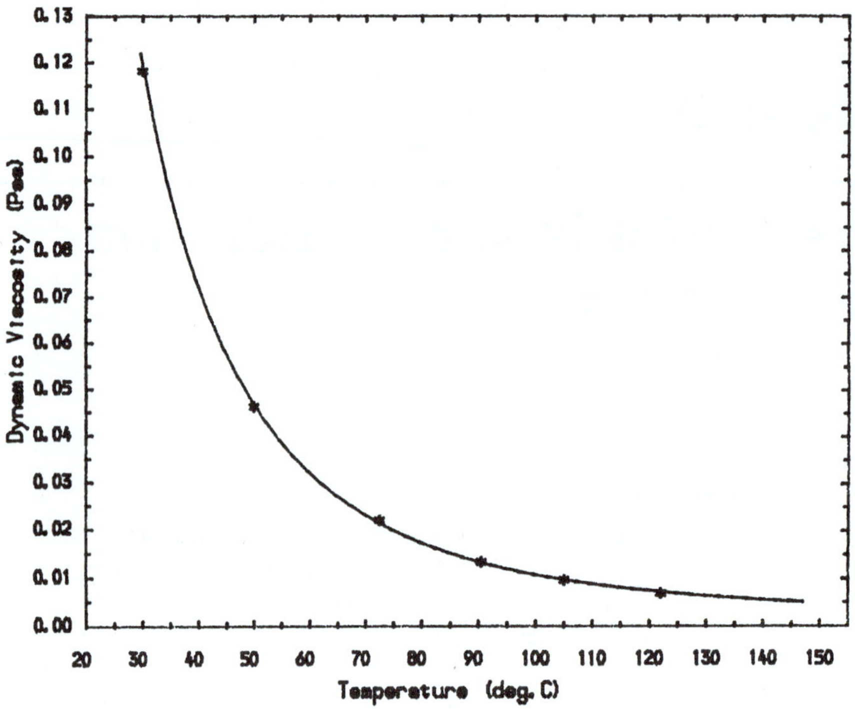

Fig. 20.1 Temperature–viscosity characteristics of the lubricant (Texaco 10W/30)

In developing the theoretical model the relationship of the valve spring force and the displacement was assumed to be linear over the whole length of the spring compression. The calibration of the tested valve spring, however, showed that the spring rate was not unique. Two spring rates were therefore adopted to simulate the relationship of the spring force and the spring compression. The spring rates were obtained from two linear regressions to ten data pairs of spring compressions and spring force measured on a stress/strain test machine. Figure 20.2 shows the calibrated relationship of the valve spring force and the spring compression. The precompression of the valve spring listed in tables was not the initial compression of the valve spring when it was assembled into the cylinder head. It was, in fact, an equivalent value to the ratio of the initial load on the valve spring at zero cam lift to the spring stiffness.

The surface roughness, the cam taper, and the radius of the curvature of the domed follower were measured from a Talysurf machine after the tests. The clearance of the follower/bore interface was measured immediately after each test.

Since the actual camshaft rotational frequency fluctuated within only one per cent of the nominal test frequencies, in the theoretical analysis the camshaft rotational frequencies were taken to be the nominal test frequencies which were 550, 850, 1150, 1450, 1750, and 2050 r/min respectively.

All the input data adopted in the theoretical analyses for simulating the test conditions at 70 and 95 °C are summarized in Tables 20.1 and 20.2 respectively.

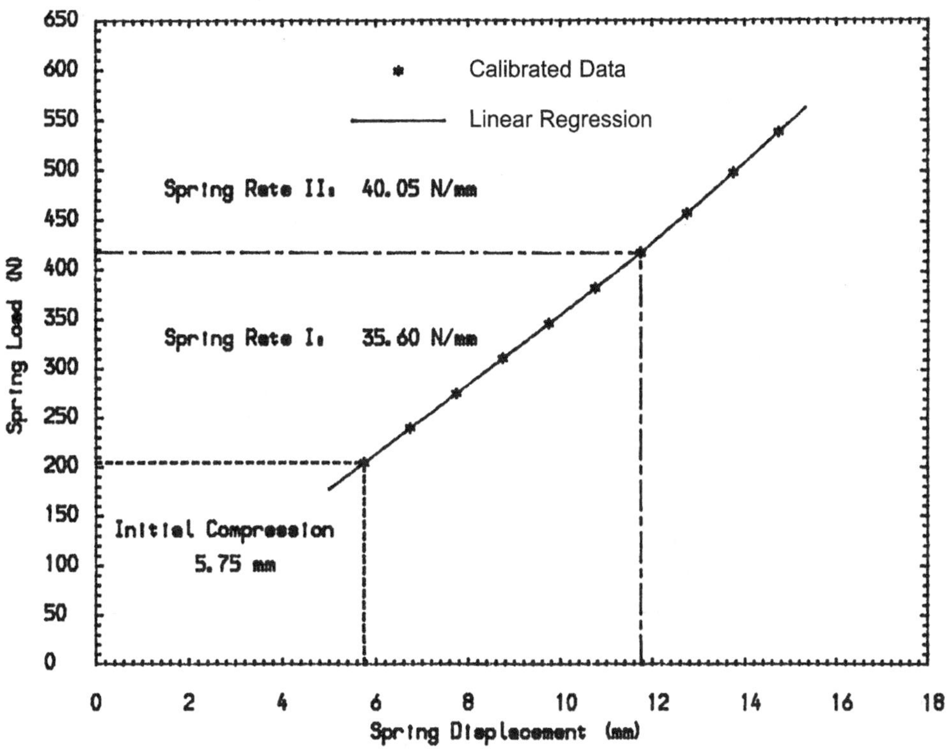

Fig. 20.2 The calibrated relationship of the spring load and displacement

20.3 The results of comparison and discussions

The accomplishment of the comprehensive experimental programme detailed in Part II has provided a vast amount of reliable data. This enables the validity of the theoretical model for predicting the tribological performance of the valve train to be examined. According to the variables measured in the tests a series of comparisons between the experimental measurements and theoretical predictions have been undertaken. This includes the average friction torque of the valve train and the average rotational frequency of the follower as a function of camshaft rotational frequency at different test temperatures. Furthermore, the experimental measurements of the instantaneous values of the total torque on camshaft, the friction torque of the valve train, and the rotational frequency of the follower have been correlated to the corresponding theoretical predictions in terms of cam angle at different test temperatures. The results of comparisons and discussions on each individual variable are detailed in the following sub-sections. For all figures presented in this chapter a solid line represents a curve fit to the experimental measurements (demonstrated with stars) and a broken line shows a curve fit to the theoretical predictions (demonstrated with crosses).

20.3.1 The average friction torque

The average friction torque of the valve train is presented in Figs 20.3 and 20.4 as a function of the camshaft rotational frequency at bulk temperatures of 70 and 95 °C. Both figures show not only the same decreasing trend of the average friction torque as

Table 20.1 The input data for theoretical analysis corresponding to the test conditions at a bulk temperature of 70 °C

Parameter	*Magnitude*	*Unit*
Cam		
Cam taper, α'	0.0573	degree
Composite roughness height of cam, σ_c	0.215×10^{-6}	m
Follower and follower bore		
Length of the follower, L_f	19.0×10^{-3}	m
Moment of inertia of the follower, J	6.8243×10^{-6}	N m s^2/rad
Mass of the follower, m_f	63.3×10^{-3}	kg
Radius of curvature of domed follower surface, R_f	19.745	m
Radial clearance between follower and bore, C_f	22.50×10^{-6}	m
Composite roughness height of follower, σ_f	0.207×10^{-6}	m
Hydraulic lash adjuster		
Inlet oil pressure to hydraulic lash adjuster, p_o	1.035×10^5 (15)	Pa (psi)
Valve spring		
Precompression of valve spring, δ	5.7501×10^{-3}	m
Stiffness of valve spring, k_{vs} (if $l_c \le \delta + 6.0$ mm)	35.601×10^3	N/m
Stiffness of valve spring, k_{vs} (if $l_c > \delta + 6.0$ mm)	40.051×10^3	N/m
Surface roughness		
Composite roughness height, $\sigma = \sqrt{\sigma_c^2 + \sigma_f^2}$	0.298×10^{-6}	m
Lubricant properties		
Lubricant Texaco 10W/30		
Lubricant viscosity at atmospheric pressure, η_o	2.3066×10^{-2}	Pa s
Test condition of valve train		
Bulk test temperature of valve train	70.0	°C
Environmental temperature of follower	62.5	°C
Camshaft rotational frequency, ω	550–2050	r/min

Table 20.2 The input data for theoretical analysis corresponding to the test conditions at a bulk temperature of 95 °C

Parameter	*Magnitude*	*Unit*
Cam		
Cam taper, α'	0.0573	degree
Composite roughness height of cam, σ_c	0.215×10^{-6}	m
Follower and follower bore		
Length of the follower, L_f	19.0×10^{-3}	m
Moment of inertia of the follower, J	6.8243×10^{-6}	N m s^2/rad
Mass of the follower, m_f	63.3×10^{-3}	kg
Radius of curvature of domed follower surface, R_f	19.745	m
Radial clearance between follower and bore, C_f	24.50×10^{-6}	m
Composite roughness height of follower, σ_f	0.207×10^{-6}	m
Hydraulic lash adjuster		
Inlet oil pressure to hydraulic lash adjuster, p_o	1.035×10^5 (15)	Pa (psi)
Valve spring		
Precompression of valve spring, δ	5.7501×10^{-3}	m
Stiffness of valve spring, k_{vs} (if $l_c \le \delta + 6.0$ mm)	35.601×10^3	N/m
Stiffness of valve spring, k_{vs} (if $l_c > \delta + 6.0$ mm)	40.051×10^3	N/m
Surface roughness		
Composite roughness height, $\sigma = \sqrt{\sigma_c^2 + \sigma_f^2}$	0.298×10^{-6}	m
Lubricant properties		
Lubricant Texaco 10W/30		
Lubricant viscosity at atmospheric pressure, η_o	1.1919×10^{-2}	Pa.s
Test condition of valve train		
Bulk test temperature of valve train	95.0	°C
Environmental temperature of follower	71.5	°C
Camshaft rotational frequency, ω	550–2050	r/min

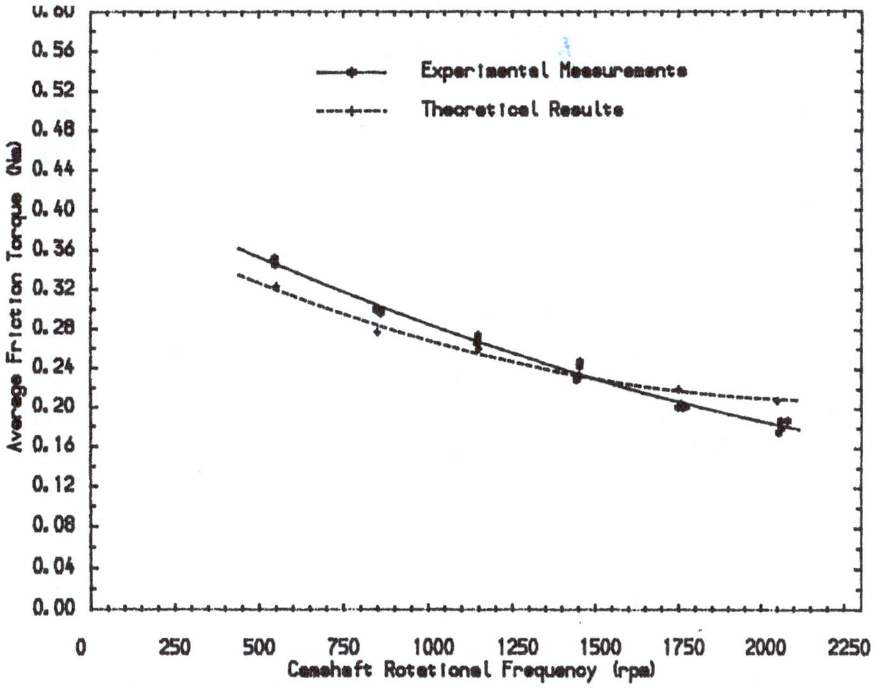

Fig. 20.3 The comparison of theoretical results and experimental measurements of the average frictional torque at different camshaft frequency (bulk temperature 70 °C)

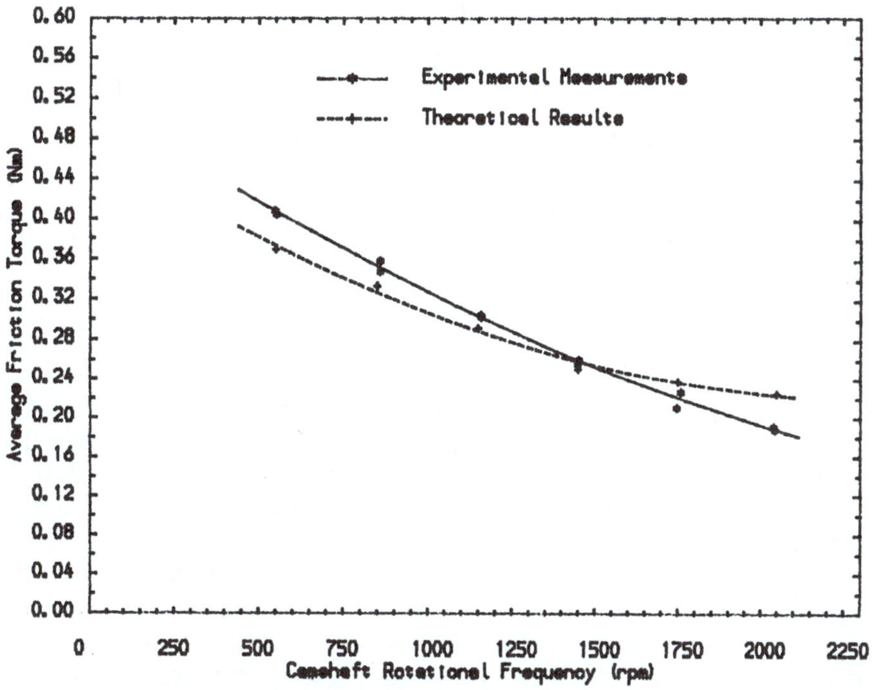

Fig. 20.4 The comparison of theoretical results and experimental measurements of the average frictional torque at different camshaft frequency (bulk temperature 95 °C)

the camshaft rotational frequency increases but also very good agreement between theoretical predictions and experimental measurements. The difference between the two curves is less than ten per cent over the whole range of test frequencies except for at the highest test frequency of 2050 r/min. The overestimation of theoretical predictions at high camshaft rotational frequency might be due to the following two reasons. First, the theoretical model developed is based upon the assumption of isothermal conditions. At high camshaft rotational frequency the sliding velocity at the cam/follower interface increases and the temperature within the conjunction increases as well. The consequence is that the actual viscosity of the lubricant decreases. The friction arising from the cam/follower interface, therefore, should be lower than that predicted by the model for the isothermal condition. Second, the dynamics of the valve train is analysed by assuming rigid components. At low camshaft rotational frequency this assumption is fairly reasonable. At high frequency, however, the predicted load on the cam will deviate from the actual cam load due to the vibration of the valve train as indicated in reference [2]. The higher the frequency the more the deviations. This will obviously influence the friction torque of the valve train. The comparison of the instantaneous torque on the camshaft will clearly demonstrate this.

20.3.2 The instantaneous torque on the camshaft

Figures 20.5 and 20.6 show the instantaneous torque acting on the camshaft rotating at different frequencies and temperatures. It can be seen that both the pattern and the value of theoretical predictions coincide remarkably well with the experimental measurements. At the highest test frequency, 2050 r/min, the measured torques fluctuate in the cam nose region as shown in Figs 20.5(f) and 20.6(f). This is caused by the vibration of the valve train system due to high rotational frequency. As discussed in the previous sub-section, the theoretical predictions which are based upon the assumption of rigid components cannot simulate the vibration of the system and consequently produce a smooth curve in the cam nose region.

20.3.3 The instantaneous friction torque on the camshaft

The instantaneous friction torques for different camshaft rotational frequencies at bulk temperatures of 70 and 95 °C are demonstrated in Figs 20.7 and 20.8. The friction torques are obtained by subtracting the geometric torque, which is estimated by the theoretical model, from the total camshaft torque of experimental measurements shown in Figs 20.5 and 20.6. The geometric torque is defined as a product of the cam load excluding the additional load due to frictions arising from follower/bore and valve/guide interfaces, and the eccentricity of the contact point of the cam and follower. The agreement between theoretical predictions and experimental measurement is remarkably good except at high frequency. As the camshaft rotational frequency exceeds 1750 r/min, the experimental measurements show some irregular fluctuations and deviate from the theoretical predictions. As indicated in the previous sub-section, the theoretical model cannot simulate the vibration of the valve train and subsequently the cam load. At high camshaft rotational frequency the influence of the vibration on the cam load becomes so great that the geometric torque which is estimated from the theoretical model deviates far from the real situation leading to substantial fluctuations in the friction torque. The difference of the friction torque at high camshaft rotational frequency between the theory and experiment is due to the lack of a proper way to evaluate the geometric torque.

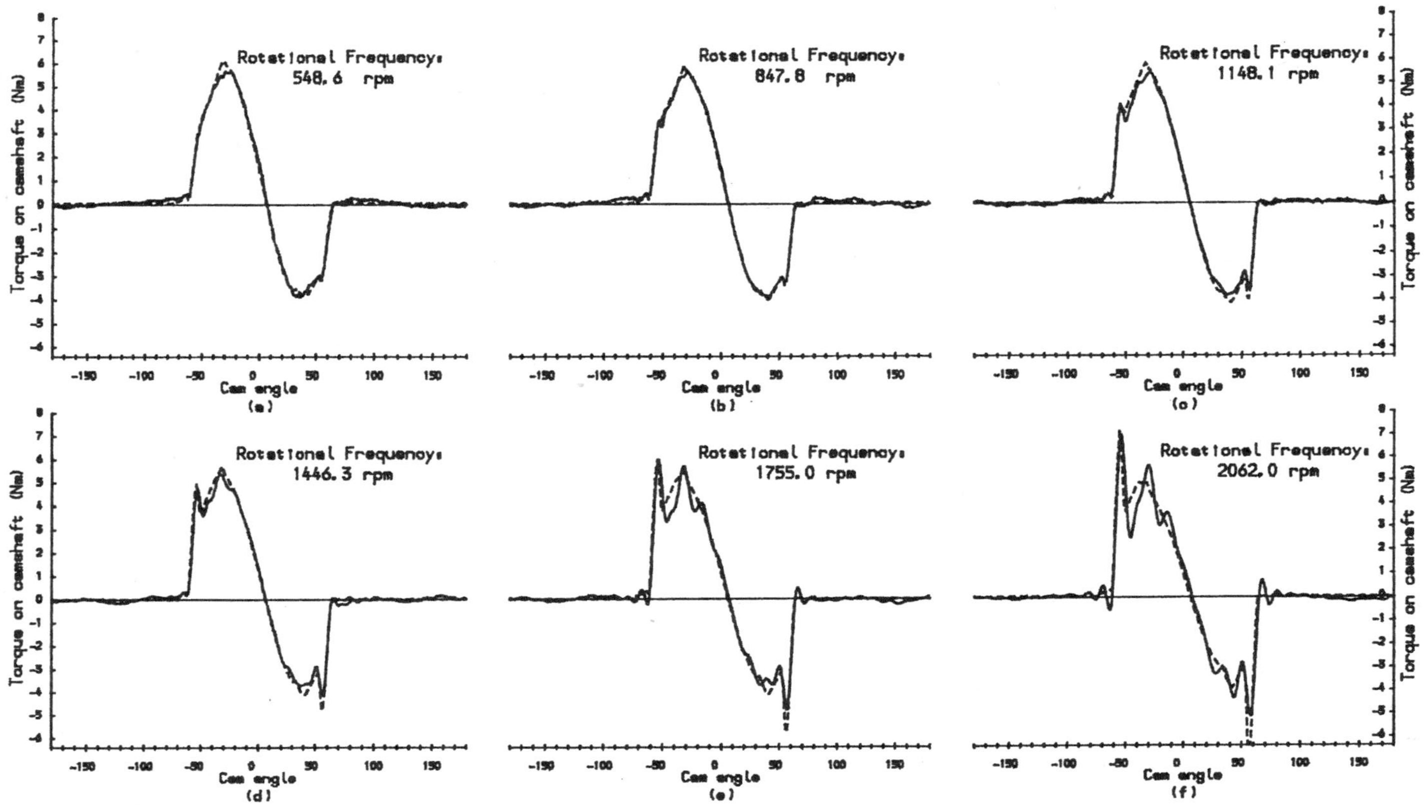

Fig. 20.5 Comparison of theoretical predictions and experimental measurements of instantaneous torque on camshaft rotating at different frequencies (lubricant Texaco 10W/30, bulk temperature 70 °C)

—— Experimental measurements, ---- Theoretical predictions

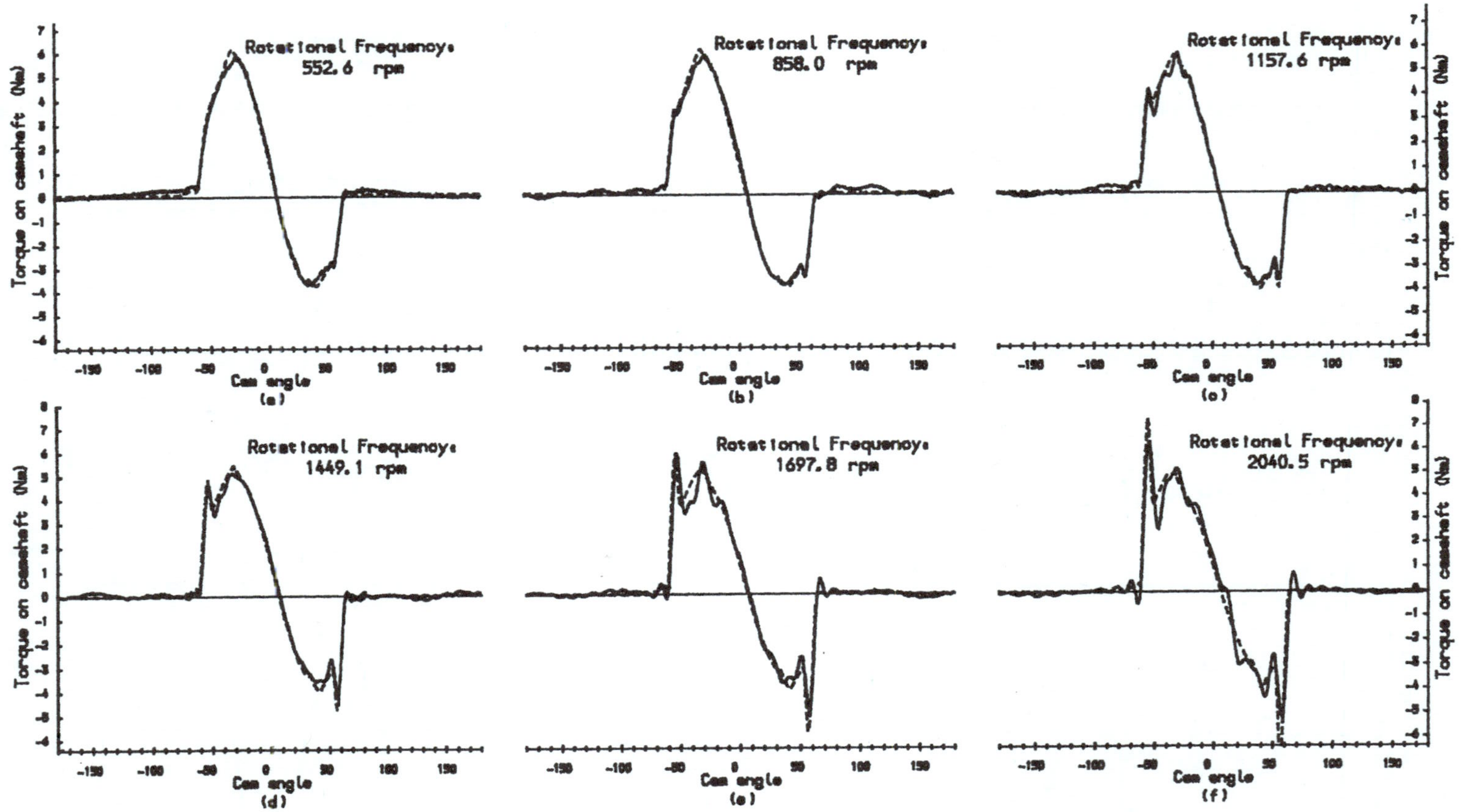

Fig. 20.6 Comparison of theoretical predictions and experimental measurements of instantaneous torque on camshaft rotating at different frequencies (lubricant Texaco 10W/30, bulk temperature 95 °C)

—— Experimental measurements, ---- Theoretical predictions

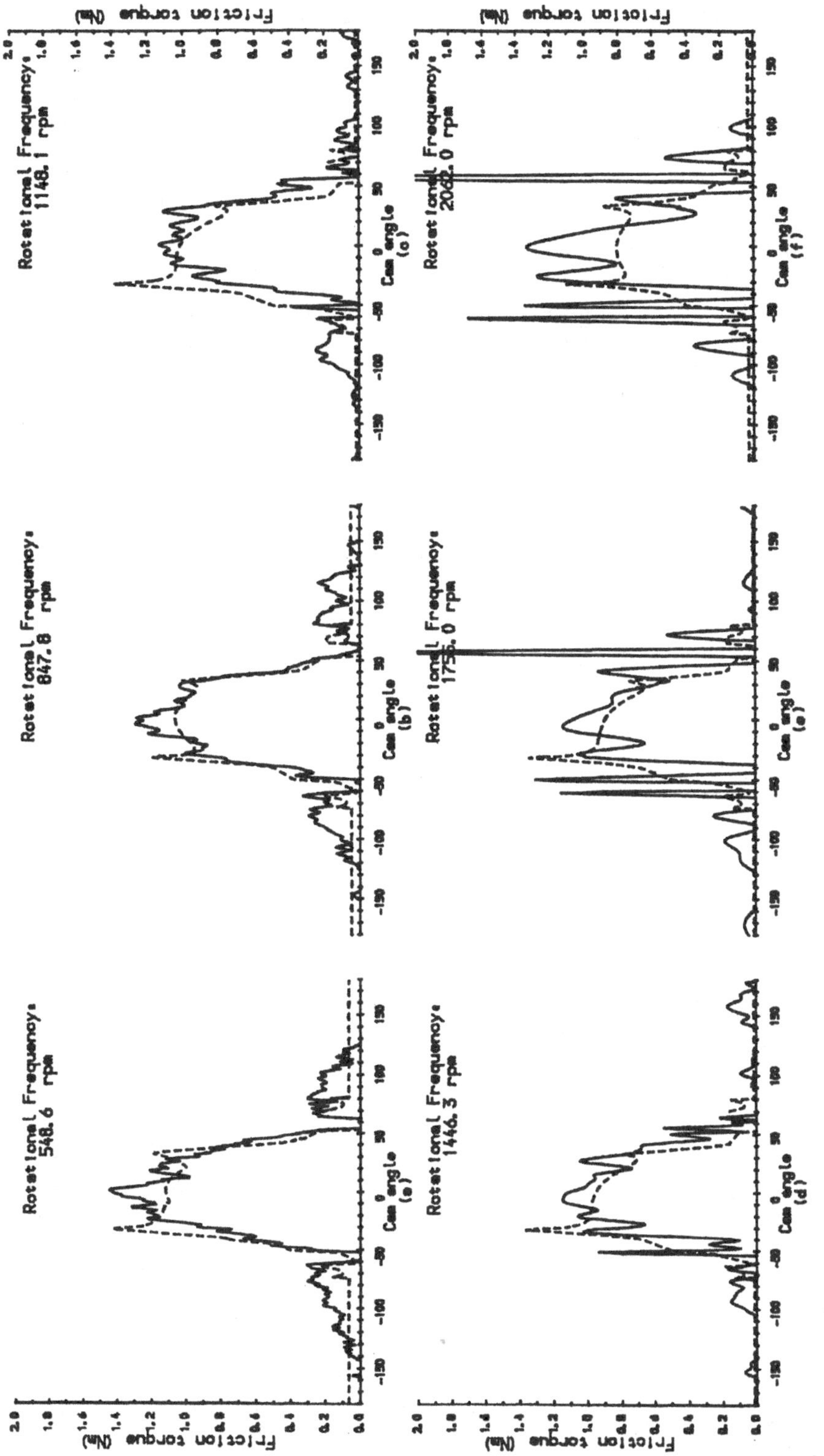

Fig. 20.7 Comparisons of theoretical predictions and experimental measurements of friction torques of the valve train at different camshaft frequencies (lubricant Texaco 10W/30, bulk temperature 70 °C)

——Experimental measurements, ----Theoretical predictions

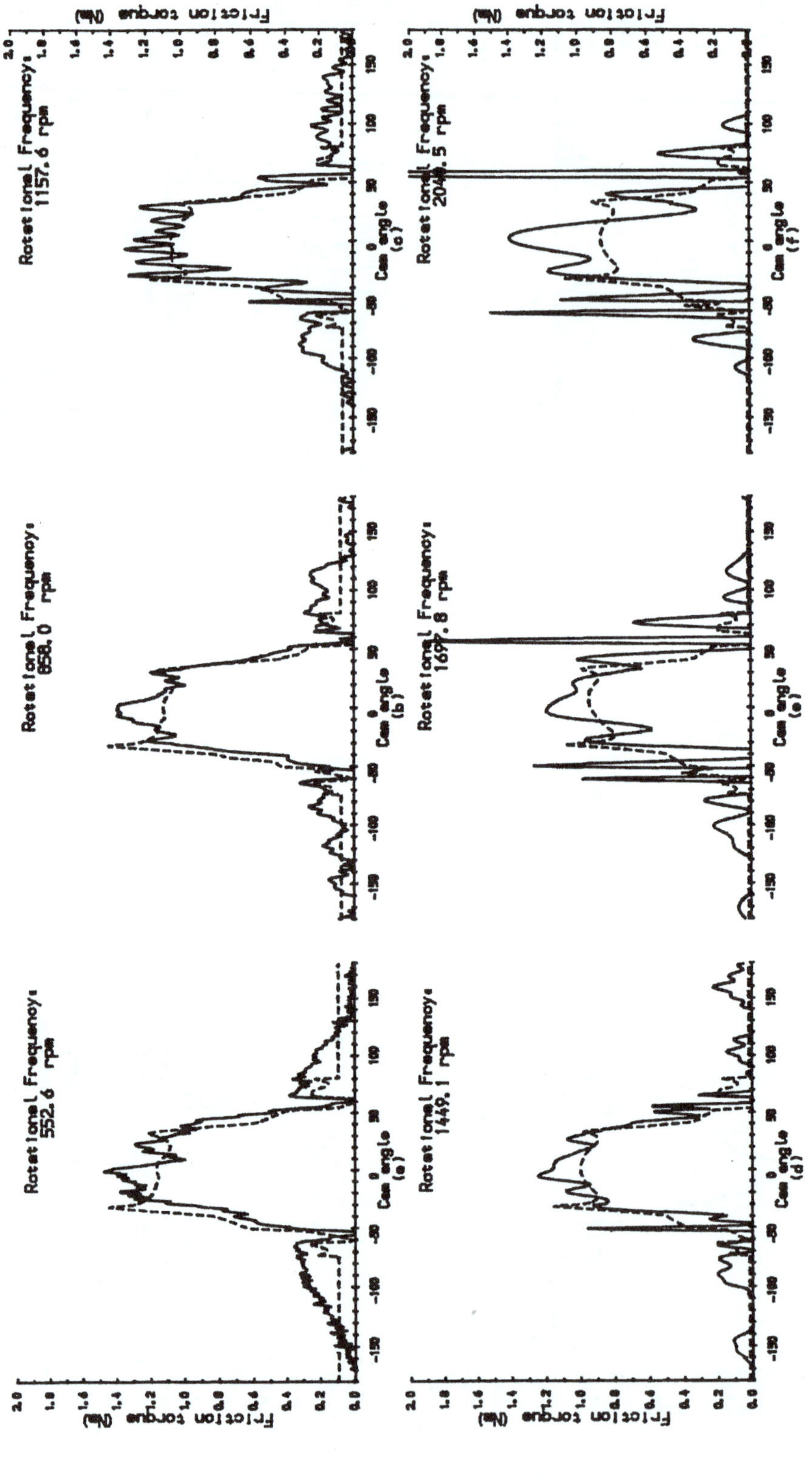

Fig. 20.8 Comparisons of theoretical predictions and experimental measurements of friction torques of the valve train at different camshaft frequencies (lubricant Texaco 10W/30, bulk temperature 95 °C)

——Experimental measurements, - - - -Theoretical predictions

On the cam base circle (between cam angle –120 and –75 degrees, and 80 and 120 degrees) the theoretical model underestimates the friction of the cam/follower interface especially at low camshaft rotational frequency. This is thought to be due to the misalignment between the camshaft centre and the centre of the cam base circle. This is also one of the reasons why the theoretical model predicts lower average friction torque than the experimental results at lower camshaft frequency.

20.3.4 The average follower rotational frequency

The average follower rotational frequencies of theoretical predictions and experimental measurements at different test temperatures are plotted in Figs 20.9 and 20.10 as a function of camshaft rotational frequency. It can be seen that the average rotational frequencies of the follower predicted by the theoretical model coincide quite well with the experimental results both qualitatively and quantitatively. The rotational frequency of the follower is quite sensitive to the lubricant viscosity and the clearance between the follower and the bore, both of which are temperature-related parameters, and other factors. It is, therefore, expected that the model over- or underestimates the follower rotational frequency under certain circumstances. In the present model the viscosity of the lubricant is taken at an average temperature of the inlet oil to the hydraulic lash adjuster and the test temperature. Considering the variation of the instantaneous rotational frequency of the follower in one cam cycle and the fluctuation of average

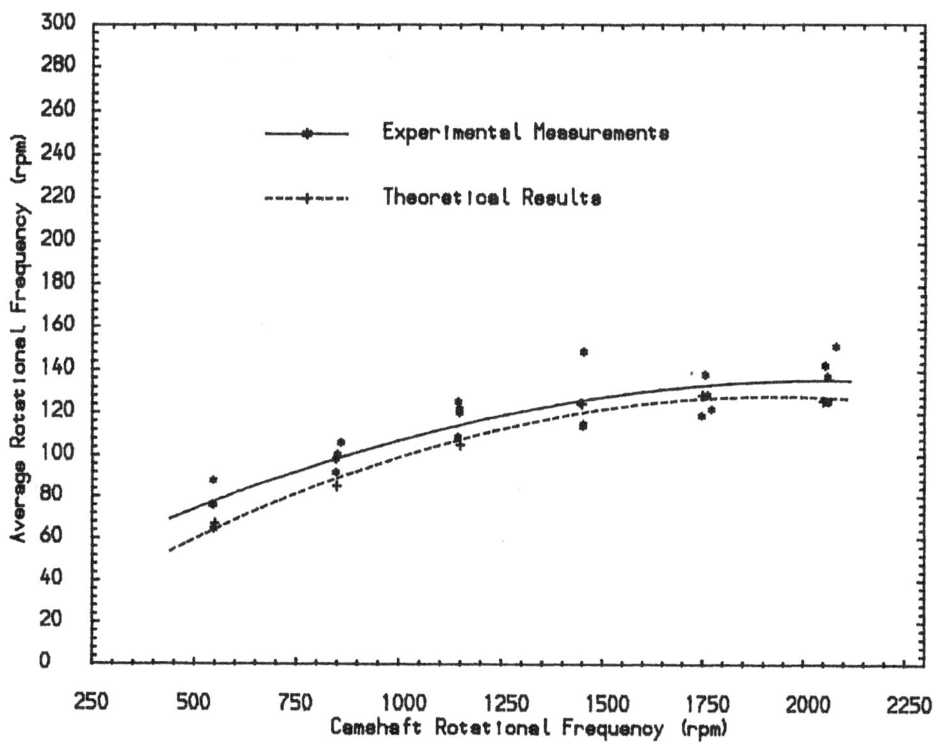

Fig. 20.9 The comparison of theoretical results and experimental measurements of the average rotational frequency of the follower (bulk temperature 70 °C)

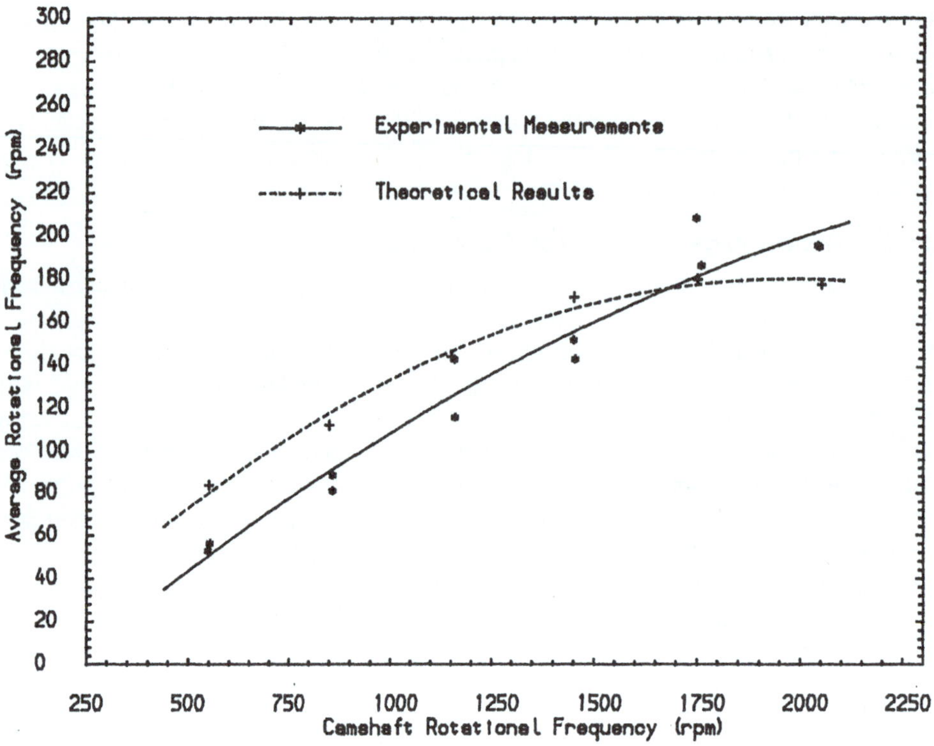

Fig. 20.10 The comparison of theoretical results and experimental measurements of the average rotational frequency of the follower (bulk temperature 95 °C)

frequency in each cam cycle as demonstrated in the second part of the report, the theoretical predictions are quite acceptable. The further comparison of theoretical and experimental results on the instantaneous follower rotational frequencies as a function of cam angle confirms the reliability of the model.

20.3.5 The instantaneous follower rotational frequency

According to the experimental evidence presented in Part II it has been known that in spite of individual differences of the magnitude of the instantaneous follower rotational frequencies, they exhibit a similar pattern in each cam cycle and repeat over several cycles. This makes it impossible to carry out a quantitative comparison on the instantaneous values for each specific test case. It is, however, still possible to test the theoretical model qualitatively to see if it is reasonable. Figures 20.11 and 20.12 show the comparison of the experimental measurements with the theoretical predictions on the instantaneous rotational frequencies of the follower. Since the measured rotational frequency is an average value within a certain period (equivalent to the period of the follower crossing one pair of black and transparent stripes) the experimental measurement is represented by a single star instead of a curve in previous figures. It is very interesting to see that under all circumstances the theoretical model predicts exactly the same characteristics that have been revealed by the experimental measurements detailed in Part II. This confirms that the physical principles and the

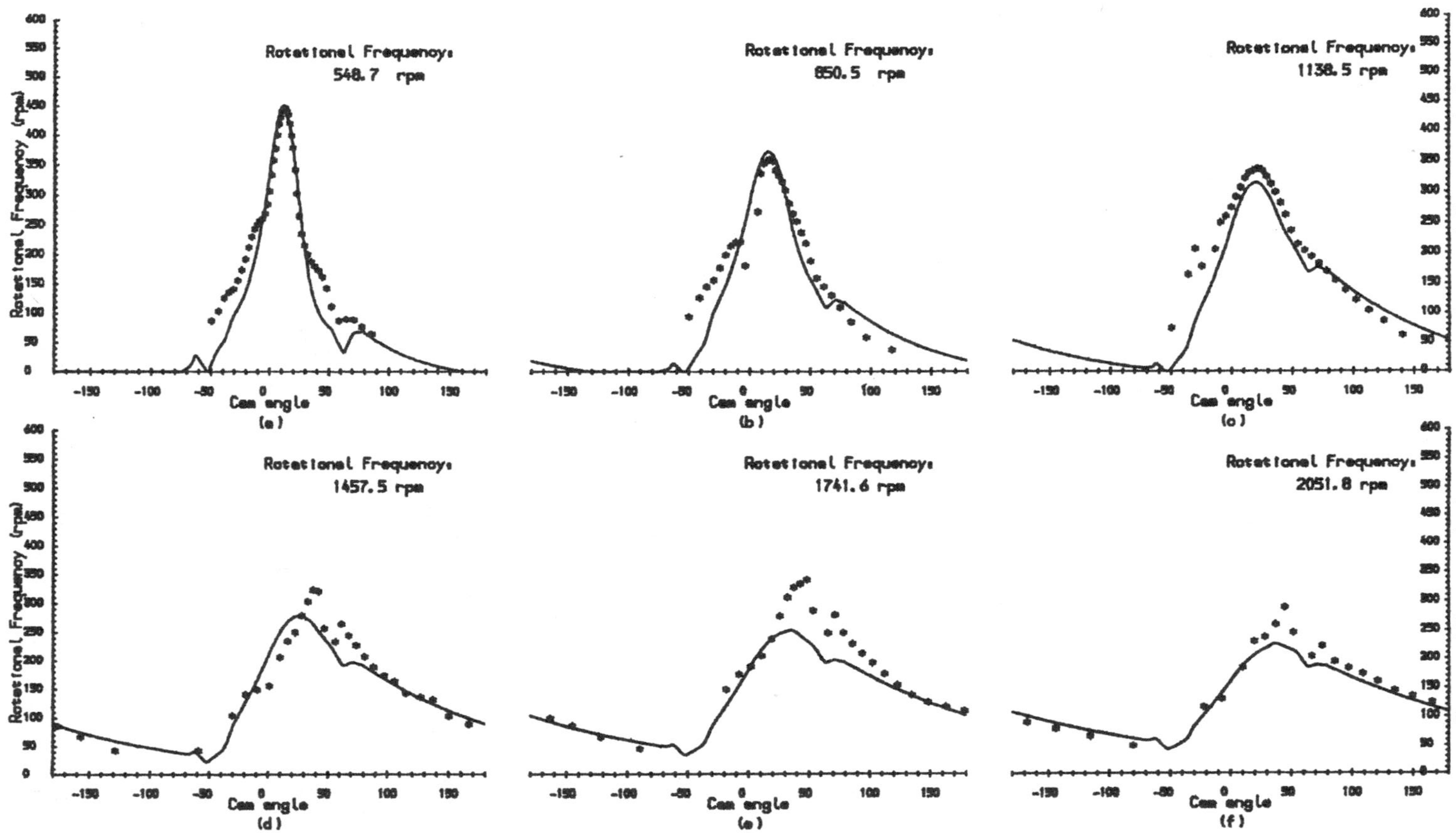

Fig. 20.11 The comparison of theoretical predictions and experimental measurements of the instantaneous rotational frequency of the follower at different camshaft rotational frequencies (bulk temperature 70 °C)

—— Theoretical predictions, * Experimental measurements

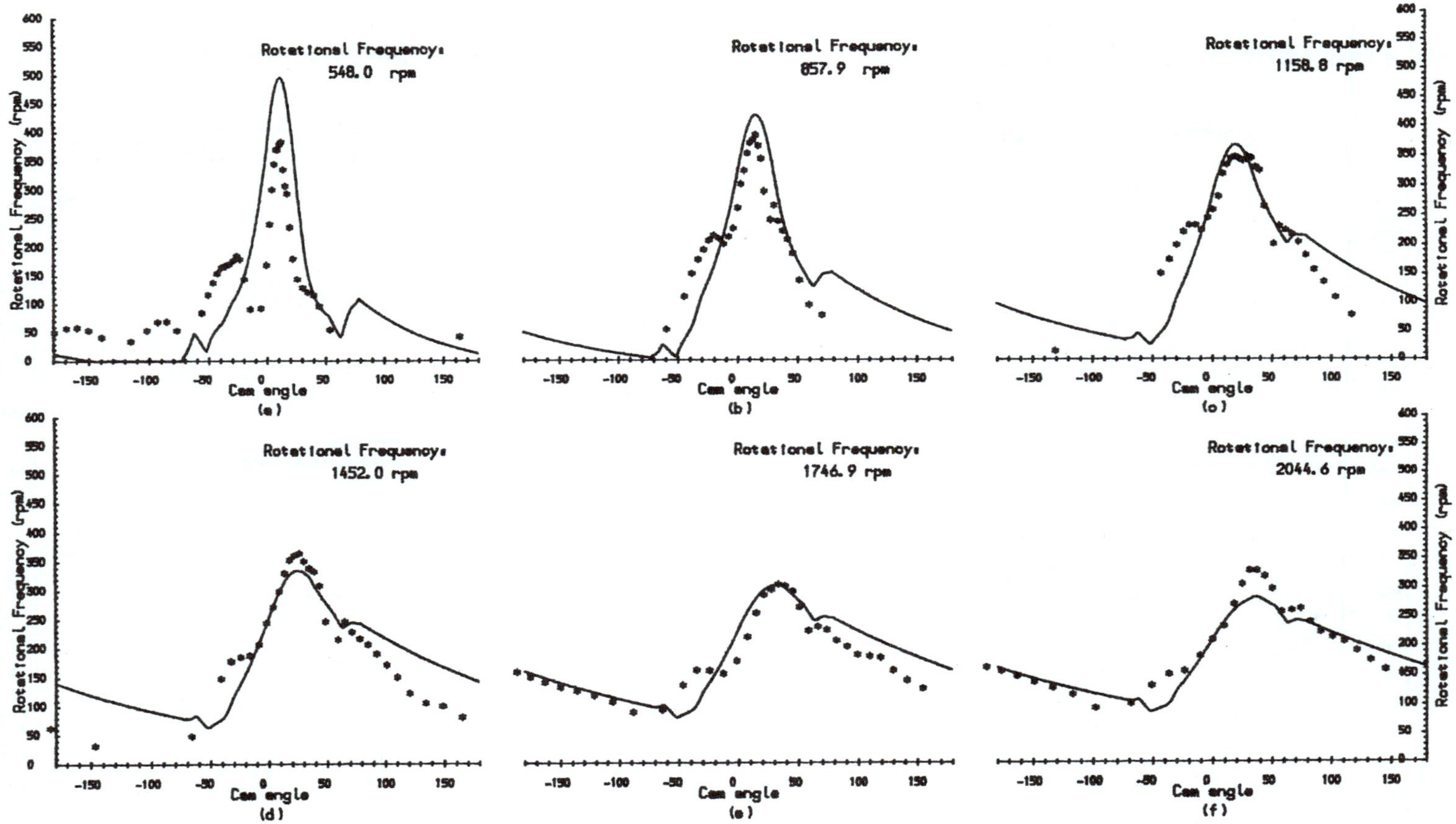

Fig. 20.12 The comparison of theoretical predictions and experimental measurements of the instantaneous rotational frequency of the follower at different camshaft rotational frequencies (bulk temperature 95 °C)

—— Theoretical predictions, * Experimental measurements

assumptions upon which the model is based are reasonable, and the major factors which influence the follower rotation have been included in the model for simulating the follower rotation.

20.4 Conclusions

A theoretical model has been developed to predict the tribological performance of a valve train having an offset taper cam and a domed follower with hydraulic lash adjuster. The model can be used to provide a lot of useful information which includes the kinematics and dynamics of the valve train, the maximum Hertzian stress, the central and minimum film thickness, the frictions arising from the interfaces of cam/follower, follower/bore, and valve stem/guide, the rotation frequency of the follower, and the performance of the hydraulic lash adjuster.

A careful and comprehensive experimental programme consisting of a series of tests at different camshaft frequencies and bulk temperatures has been undertaken to investigate the tribological performance of a valve train and to validate the theoretical model developed for predicting the tribological performance. Several important variables including the friction torque, power loss, follower rotational frequency, and temperatures in the valve stem guide and in the oil inlet to the hydraulic lash adjuster have been monitored in each test.

A multi-aspect comparison between the theoretical predictions and experimental measurements has been undertaken. This includes both qualitative and quantitative comparisons of the average and instantaneous values of the friction torque and the rotational frequency of the follower. The excellent agreement between the theoretical predictions and the experimental measurements in many aspects shows that the theoretical model provides a strong basis for predicting the tribological characteristics of the valve train having an offset taper cam and a domed follower with hydraulic lash adjuster. This gives great confidence in the model developed in the current research project and the future application to the valve train design of internal combustion engines.

20.5 References

1. **Zhu, G.** (1988) A theoretical and experimental study of the tribology of a cam and follower. PhD thesis, Mechanical Engineering Department, Leeds University, UK, May.
2. **Heath, A. R.** (1988) Valve train design for multivalve automotive gasoline engines. SAE Paper 885133.

Chapter 21

Overall Conclusions

The work presented in this book has been concerned with both theoretical and experimental studies of the tribological performance of an automobile valve train having an offset taper cam and a domed follower with hydraulic lash adjuster. The work has included three main parts:

(a) the theoretical formulation;
(b) the experimental investigation; and
(c) the correlation of theory and experiments.

First, a sophisticated theoretical model has been developed to predict the tribological performance of the valve train. The development of the model is to provide a useful tool for the tribological design of the engine valve train. Several important tribological characteristics of the valve train can be predicted by this model. These include the information associated with the cam/follower interface:

(a) the kinematics and dynamics;
(b) the maximum Hertzian stress;
(c) the central and minimum lubricant film thickness;
(d) the load and friction due to asperity contact; and
(e) the friction and power loss.

In addition, the model can estimate the instantaneous and the average rotational frequency of the follower. Furthermore, the performance of the hydraulic lash adjuster can be predicted by the current theoretical model. The friction arising from the other interfaces of the valve train components, such as tappet/bore and valve stem/guide interfaces, can also be estimated by the model.

Second, a careful and comprehensive experimental programme consisting of a series of tests at different camshaft rotational frequencies and different bulk temperatures has been undertaken. This is to investigate the tribological performance of the valve train and to validate the theoretical model. Several important variables including the friction torque, the power loss, the follower rotational frequency, and the temperatures of the lubricant supply, inside the test cell and in the valve guide, have been monitored in each test. Some sophisticated techniques for data acquisition and data processing including a PC-based data acquisition system and a follower rotational frequency monitoring system have been developed. The accomplishment of the comprehensive experimental programme has provided a vast amount of reliable data. This enables the validity of the theoretical model to be examined.

Finally, in order to validate the theoretical model developed to predict the tribological performances of the valve train, the experimental measurements have been correlated with the theoretical predictions which simulate the test conditions of the valve train. The agreement between the experimental measurements and theoretical predictions is excellent.

The overall conclusions which have emerged from the study in this report are now summarized.

1. The remarkable consistency and repeatability of the results demonstrates that the experimental measurements are very reliable and the facilities and the technology developed in the current experimental study can readily be used to investigate the tribological performance of various configurations of direct-acting valve trains.

2. Although the follower is free to rotate there is still a substantial amount of sliding between the cam and follower. This makes the major contribution to the friction of the valve train. The average friction torque of the valve train decreases as the camshaft rotational frequency increases. An increase of the bulk temperature leads to an increase of the average friction torque of the valve train.

3. The average follower rotational frequency increases with increase of the camshaft rotational frequency. An increase of the bulk temperature results in an increase of the average rotational frequency of the follower.

4. The maximum instantaneous rotational frequency of the follower decreases as the camshaft rotational frequency increases, while the rotational frequency of the follower on the cam base circle increases. The higher the camshaft rotational frequency the more evenly the follower rotates.

5. The inspection of the cam and the follower surface before and after the tests indicates that the actual contact centre on the base circle is in the vicinity of the centre of the follower.

6. The excellent agreement between the theoretical predictions and experimental measurements demonstrates that the theoretical model developed in the present study is reliable for future application to predict the tribological performance of an automobile valve train. This gives great confidence in dealing with novel and alternative designs with the theoretical model.

Index

Printed in the USA/Agawam, MA
May 19, 2026

905303.003